Python for Data Science

The ultimate guide for beginners. Machine learning tools, concepts and introduction. Python programming crash course.

Tony F. Charles

© **Copyright 2019 - All rights reserved.**

The content contained within this book may not be reproduced, duplicated or transmitted without direct written permission from the author or the publisher.

Under no circumstances will any blame or legal responsibility be held against the publisher, or author, for any damages, reparation, or monetary loss due to the information contained within this book, either directly or indirectly.

Legal Notice:
This book is copyright protected. It is only for personal use. You cannot amend, distribute, sell, use, quote or paraphrase any part, or the content within this book, without the consent of the author or publisher.

Disclaimer Notice:
Please note the information contained within this document is for educational and entertainment purposes only. All effort has been executed to present accurate, up to date, reliable, complete information. No warranties of any kind are declared or implied. Readers acknowledge that the author is not engaging in the rendering of legal, financial, medical or professional advice. The content within this book has been derived from various sources. Please consult a licensed professional before attempting any techniques outlined in this book.

By reading this document, the reader agrees that under no circumstances is the author responsible for any losses, direct or indirect, that are incurred as a result of the use of information contained within this document, including, but not limited to, errors, omissions, or inaccuracies.

Table of Contents

Introduction 1
Chapter 1: Getting Started 4
    Python Setup 9
    Python Scientific Distributions 11
        Anaconda 12
        Canopy 14
        WinPython 15
    Virtual Environments 15
    Data Science Packages 19
        Jupyter 22
    Summary 24
Chapter 2: Data Munging with Pandas 25
    The Process 27
    Importing Datasets 28
    Data Preprocessing 32
    Data Selection 36
    Summary 38
Chapter 3: The Data Science Pipeline 39
    Exploring Data 39
        New Features 42
    Dimensionality Reduction 45
        Covariance Matrix 45
        Principal Component Analysis 48
        Latent Factor Analysis 50

Outlier Detection 51

Summary...53

Chapter 4: Supervised Learning Algorithms 55

Regression..56

Linear Regression 57

Logistic Regression 59

Naive Bayes Classifier.. 60

K-Nearest Neighbors 63

Support Vector Machines 67

Summary...72

Chapter 5: Decision Trees 73

Implementing Decision Trees..73

Classification and Regression Trees............................... 77

The Overfitting Problem 78

Pruning..80

Decision Tree Implementation 82

Summary...87

Chapter 6: Unsupervised Learning with K-means Clustering..88

K-means Clustering 89

Summary...95

Chapter 7: Neural Networks 96

Neural Network Structures...97

The Feedforward Neural Network 99

Backpropagation.. 100

Recurrent Neural Networks 102

The Restricted Boltzmann Machine............................ 103

Restricted Boltzmann Machine in Practice 105

Summary 116
Chapter 8: Big Data **118**
　The Challenge 119
　Applications in the Real World 123
　Analyzing Big Data 126
　Summary 128
Conclusion **129**
Bibliography **130**

Introduction

You have taken the first step on your path to learning data science by purchasing *Python for Data Science: The ultimate python guide for beginners. Python machine learning tools, concepts and introduction. Python programming crash course.* Allow yourself to be guided in the field of data science step by step from the very beginning with the help of this book. The purpose of *Python for Data Science* is to educate the complete beginner on the fundamental topics of data science.

Studying this field can be exhausting, difficult and even intimidating to some, especially those with limited programming knowledge. Fortunately, the step by step approach of this book will teach you the theory as well as the practical skills you need to implement learning algorithms and techniques on your own in no time. All you need is a desire to learn and a good instruction manual with a practical approach towards study.

This book includes the following topics:

1. In the first chapter you learn how to setup your Python work environment by creating a virtual environment where you will install a scientific distribution that includes all of the tools you need. Keep in mind that you will be working only with free tools and open source

datasets, so you won't have to pay a dime.

2. Once your toolkit is ready, the next two chapters will teach you about the importance of data pre-processing and exploration. Do not rush through these sections because the more time and effort you invest in preparing your data, the better results you will obtain.

3. Once your data is prepared, you will begin exploring a variety of supervised and unsupervised learning algorithms and techniques. You will learn about linear regression, the Naïve Bayes classifier, the K-nearest neighbor algorithm, support vector machines, decision trees, K-means clustering and more.

4. After learning the fundamental learning techniques you will start exploring the power of neural networks. These structures are some of the most powerful data analysis techniques in data science, and this chapter will introduce you to the most popular ones. You will learn about a variety of neural networks such as the single layer and multilayer perceptron, as well as the Restricted Boltzmann Machine.

5. Finally, you will also be introduced to the field of Big Data, which is a subfield of data science that processes vast volumes of data. You will learn why this subfield is the future and what real world applications exist for it. Furthermore, you will gain an idea about which supervised and unsupervised learning algorithms can be used when handling Big Data.

Take note that in order to learn fast and benefit from this guide, you need to practice along with it. It is very tempting to start reading and not let go of the book, however, doing things that way means you won't learn much. The book is divided in such a way that it allows you to go section by section, bit by bit, and make your way up the ladder until you understand the fundamentals of data science. And make sure to practice every concept and technique!

Chapter 1: Getting Started

Data science is a relatively new discipline that has entered the market only recently. Even though this field has existed for decades, it was only studied and researched mostly on a theoretical level by mathematicians and specialists from various technical fields. Data science includes a number of components that can be studied individually as well, such as machine learning, computational linguistics, data analysis, linear algebra and many more. From these elements alone we can conclude that data science is in fact a seemingly chaotic mixture of mathematics, programming, and data communication. However, this combination is one of the main reasons why data science is so widely used for a variety of purposes. It is a versatile field. This versatility, however, leads to a number of different specialties destined for certain areas of expertise within data science.

But before we discuss what data science is in detail, you should understand what it isn't. Many beginners get the wrong idea about this field and they end up making mistakes that can set them back on their journey.

First of all, know that data science is not equal to software development or engineering. Your job as a future data scientist will not be developing products, systems, or features. Secondly, data science is not about graphical work, or visualization. You will not develop tools focusing on cool visuals or graphical elements. Thirdly, data science does not involve scientific work in the sense of academia. The field is determined by what the industry or market requires from the data scientist. Finally, there is a false image of data science involving mostly statistics. The field of statistical analysis and anything related to that subject can be part of data science and having that kind of knowledge can prove to be useful, however, there is far more to it than that. Statistical knowledge alone will not qualify you to become a data scientist.

With that in mind, let's now discuss what data science actually is. This field involves programming skills, statistical understanding, visualization concepts, and finally a business sense. This last component may often be harder to acquire, however, it is just as important as the others, if not more so. When we refer to a business sense we are discussing the ability to answer any possible business questions with all the available data you are provided with. You need to be capable of connecting the dots by using only the carefully selected data from a world full of data. Essentially, the data scientist is the one who connects the world of business with the technical world of data.

To keep it simple for now, data science can be compared to cooking. You start preparing your ingredients first by cleaning the vegetables, the meat, and chop everything to the desired size. In the same way you start the data science process through data munging, which involves data cleansing, debugging, extraction, loading and so on. Once this step is completed, you start cooking your ingredients, step by step, timing every process until the raw food can be eaten. This step corresponds to data exploration and various processes such as feature reduction or construction. Now the meal is ready to be served and you arrange it nicely on your plate in order for it to be nice and presentable, and you serve everything in a specific sequence. In data science this final step is the part where you display your data mining results, perhaps with the aid of graphical visualization in order to make it easy for your business users to understand the data reports.

Now that you have an idea about the ingredients behind the cooking process, you need to obtain the right set of tools. Developing a professional toolkit from the beginning is a crucial step. In the field of data science you will always be learning new techniques and add to your toolkit. However, there is one tool that serves as the foundation for everything else, and that is Python.

Python is one of the best multi-purpose tools in data science, as well as other technical specializations. In this case, it is particularly powerful when it comes to data munging, as well as data mining, especially with the help of packages and libraries such as Scikit-learn. However, it's worth mentioning that sometimes R is used as the alternative, however, it is somewhat lacking when it comes to data munging. It is often used as a way to prototype your data science problem because of its usefulness when undergoing data exploration or testing algorithms under different conditions.

In this book, we are going to focus on Python due to its simplicity, power, versatility and ability to be extended with a variety of open source libraries and packages. This programming language is used in many fields, as it's becoming increasingly impossible to work with computers without hearing about Python. Some people consider it nothing but a trend for beginners, however, that's not the case as it is used in everything from software development to machine learning and data science by everyone whether a complete novice or professional.

Python's humble beginnings are found in 1991 when Guido van Rossum, a big fan of the famous Monty Python comedy group, designed it. Speaking of comedy, if you are a fan yourself, keep your eyes open as you go through the documentation for this language and you'll find several fun Easter eggs. With that bit of trivia out of the way, here are some of the main characteristics that make Python the optimal choice for studying data science:

1. Python is powerful: This programming language can equip you with every tool you need for every step you need to perform and it won't cost you a thing. Being an open source product has its advantages. Developers from around the world continue to improve and refine the language, especially with the addition of various packages, libraries, and modules. Python is more

powerful than ever before because of how much its usability can be extended.

2. Python is versatile and user friendly: This feature is tied to the first one in many ways. First of all, Python is one of the most beginner-friendly programming languages you can learn and that is why it is often recommended for aspiring data scientists and machine learners. All you need is a basic grasp on the language and then you can focus on the data science concepts and techniques. Secondly, the ability to extend the language makes it easier for you to work with algorithms, especially if you are a complete beginner.

3. Python is easy to integrate: Many other tools and even programming languages can be used in combination with Python. For instance, if you know how to code in C or C++, you can use that language for the bulk of your programming tasks and then use Python as the central hub which connects all your tools to your code.

4. Python is compatible with many platforms: Portability is sometimes a problem, especially when you work on one system to achieve your goals and then you have to move your entire work to a different system. This is where Python excels. The code and the tools you work with using Python will automatically run on any operating system, whether it's Windows, Linux or Mac.

To summarize, Python is powerful, easy to learn and even master, and can be integrated with a variety of tools, programming languages as well as operating systems. With that being said, let's dive in and start installing Python, as well as all the other tools we will need.

Python Setup

In this section we are going to discuss the installation process for Python, as well as a number of its extensions. If you are already familiar with the programming language, you can skip the setup and continue reading about libraries and scientific distributions. If this is the first time you are installing it, however, you should know that Python comes in two different versions. You have the option to install Python 2 or Python 3. As you can probably tell, version 2 is the older one, however, it is still in use and many data scientists choose to stick with it for the time being. Read up on the documentation and perhaps check what the online Python community has to say as well and then decide which version to use. Some tools and extensions work only with Python 2, so keep that in mind as well.

In this book we are going to focus on the up to date Python 3 because this version is the future and it will continue receiving support for quite some time. Whatever your decision is, please stick to one of them all the way in order to avoid any compatibility issues. If you write your code in Python 2 and then you execute it in a system that runs with Python 3, you might encounter serious errors.

With that in mind, head to Python's official webpage at www.python.org and download your chosen version of Python for the operating system your computer is using. Keep in mind that for the purpose of data science, installing Python alone is not enough. Therefore, you have two options. You can either install everything manually and select each element and tool as needed, or install a Python scientific distribution, which essentially is a package that contains everything you need. Nonetheless, we are going to discuss both methods because having full control over what is needed for your project is important.

Once you downloaded Python, simply follow the instructions to perform the basic setup. Now, you need to also install a variety of packages, otherwise the language alone with its base components isn't that useful for data science. To perform this step manually, you need to use a package manager. Fortunately, Python comes with one already installed and it's called "pip". Open you command terminal and type pip in order to launch the manager. This package manager will allow you to install and uninstall any component whenever you wish. You can even perform updates to every single package, or only to certain ones. You even have the option of rolling back to a different version if the update causes some kind of conflicts within your project. With that being said, let's install a package by typing this command:

pip install < myPackage >

That's it! You only need to know the name of the package you want to install and use its name as a keyword. Updating and removing the packages is done the same way. Simply replace the "install" keyword with the appropriate one for each command.

Now that you have an idea about the basic installation process, let's discuss scientific distributions. Installing and maintaining dozens of packages and modules by hand can be quite tedious, so let's automate it.

Python Scientific Distributions

Creating your Python working environment can be time consuming and can lead to compatibility errors if you go through the process manually. Some tools don't work well together, or they require the installation of certain dependencies you aren't aware of. It is good to be able to customize your Python installation, however, as a beginner you might want to stick to a Python distribution instead.

Save your time and effort to practice data science and machine learning concepts and techniques instead of fiddling with various packages. A scientific distribution will allow you to automate the entire installation process, finish it within minutes and then jump straight into practical data science. There are different scientific distributions out there and they each contain a certain set of tools designed for various types of projects. However, they all have in common the fact that they contain a number of powerful Python libraries, specifically designed with data science in mind, as well as an Integrated Development Environment and other tools to get you started as soon as possible. With that being said, let's discuss the most important distributions that you should download and explore.

Anaconda

One of the most popular scientific distributions you can find is Anaconda. In a way it completes Python by extending it with around 200 pre-installed packages that include the ones data scientists use the most, such as NumPy, Pandas, Matplotlib and Scikit-learn. Furthermore, Anaconda can run on any computer no matter which operating system and version you're running and can even be installed next to other distributions without causing any conflicts.

This scientific distribution contains everything you need, especially in the beginning of your journey. It offers you all the tools needed to perform data analyses and explorations, as well as mass data processing and computing. In addition, it comes with its own package manager, which allows you to tweak the packages yourself or maintain them manually if you prefer.

However, the most attractive part about using Anaconda is the fact that it offers you all of that for free, while other distributions might involve certain fees.

If you decided to go with Anaconda, you will need to get accustomed to its internal package manager which is accessed with the keyword "conda". The manager works just like pip and is used to install, delete, or update any or all of your packages. Furthermore, it can also be used to setup a virtual environment (we'll discuss it soon). Now, let's check to see if we are running the most recent version of conda before we do anything else. Perform the verification and update by simply using the update keyword like so:

conda update conda

Now, in order to manually install certain packages, because the distribution might not contain quite everything you need, you need to use the install keyword just like you did when testing out the pip package manager. You can even install several packages at the same time by simply listing them all in a sequence. Here's how that looks like in the command console:

conda install < package1 > < package2 > < package3 > < package100 >

Next, you can update the packages with the update function. The easiest way to do this is to simply update everything at once by adding the "—all" keyword. Here's how:

conda update –all

Lastly, you might want to remove some of the packages you don't need, or if you installed one that causes some kind of conflicts with another package. Simply use the "remove" keyword to instruct the manager to remove the desired package. It's that simple.

Conda is a very intuitive package manager and it is easy to use. If you discover that you prefer it over other managers like pip, then you can even install it separately to use it with any other scientific distributions.

Now, let's explore another scientific distribution called Canopy.

Canopy

This distribution is a popular alternative to Anaconda because its aim is to provide data scientists and analysts with all the tools they need to do their job. It also contains approximately 200 packages, including Pandas and Matplotlib. Make sure to check the distribution's documentation to see all the other packages it contains because they are too numerous to list here.

In many ways Canopy is similar to Anaconda. They both have many packages in common, however, without paying a fee you will only have basic access to Canopy. For starters that's enough, however, if you ever need its advanced features you need to install the main version of the distribution.

WinPython

This distribution is aimed at Windows users, as the name suggests. Essentially, it offers you the same features as Anaconda and Canopy, however, while the others have paid versions with more advanced features, this distribution is entirely community driven and therefore free. WinPython is an open source tool that you are free to use however you want and you can even modify it yourself if you have the required skills to do so.

The main drawback of course is the fact that it can be installed only on a Windows machine, but if you do use one, you should consider it because it even allows you to install several versions of it on the same computer. In addition it also offers you a free IDE.

Virtual Environments

Now that you have an idea about scientific distributions, let's take a sidestep and discuss the power of virtual environments before exploring various Python libraries and other tools you need.

Virtual environments are one of the most useful tools when practicing something new, or when you want to be able to perform multiple tasks in different environments. Keep in mind that you are normally restricted by your operating system or by the version of Python which you currently installed on your system. No matter the chosen installation method, it is difficult to install different versions of the same tools without risking potential software conflicts and errors. However, if you take the virtual environment route, you will have a number of advantages and no significant disadvantages.

One of the biggest benefits of working through a virtual environment is the fact that it can be your testing ground. This means that you can dedicate an area of your system to experimentation and learning without any consequences. You can try any variety of libraries, modules, different versions of your tools and so on. Furthermore, you can learn and practice data science concepts and techniques without being worried that you will damage your system in some way. Eliminate the stress of having to reinstall your operating system or fix complicated errors by learning to work through a virtual environment. The worst case involves resetting the environment, which means you don't need to Google obscure error codes or try to fix your operating system because you messed something up.

Studying aside, another reason why so many choose to work through a virtual environment is because it allows them to have multiple Python and data science distributions installed at the same time. In addition, you can also install multiple versions of your operating system to see how your project behaves. You can even install different operating systems altogether on the same machine without affecting your main system. Some of the packages and modules work only with certain versions of Python for instance. Therefore you can use a virtual environment for Python version 2 and another one for Python version 3. Create unique virtual environments that suit your needs precisely.

Finally, a virtual environment will help you verify the replicability of your project. Sometimes you need to make sure that whatever task you performed or project you worked on can be replicated on a different operating system or with a different Python version. By installing different versions of your tools or the operating system in separate virtual environments you can test your prototype and see how it performs.

Now that you understand what virtual environments are for, let's install one with pip:

pip install virtualenv

Installation is quite straightforward, however, you will need to make some considerations at the end of the setup. For instance, you will need to decide which version of Python your virtual environment should use. The default version is the one from which virtualenv is installed. Next, you need to consider your packages. With each virtual environment installation, all of the packages will also be installed. This is a good thing because it's one less task you need to perform, however, when you setup multiple virtual environments each one of them will install the same packages once again. This is a waste of system resources, as well as time. There is no need to install the same thing more than once, therefore you need to issue a command for the virtual environment to use the files it already has access to instead of installing a new set. Finally, you need to consider whether you need the option to relocate your virtual environment to a different system that perhaps even uses a different version of Python. To make your virtual environment relocation-friendly, you need to specify that you want the scripts to work no matter the path.

Once you choose the correct options for your situation, you can finally create and launch the environment. Type this line to generate the new environment:

virtualenv myEnv

A new virtual environment will be created together with a directory named "myEnv" where it will reside. Now let's launch the environment with these commands:

cd myEnv

activate

The environment is now ready and we can install all of the tools we need as before. The environment acts exactly the same as your system so everything you did so far you can do in any virtual environment you create.

Data Science Packages

So far we kept mentioning how powerful Python is because of the wide variety of packages, libraries, and extensions that we have access to specifically for data science and analytics. As a beginner data scientist you will need to understand these packages because they represent your toolkit without which your knowledge can't be easily applied.

In this section we are going to briefly explore the most important Python packages and tools you will be working with throughout this book. Keep in mind that these tools have been selected because they are industry standard, polished, and frequently updated, and most importantly well-documented.

1. Scikit-learn: This is one of the most important tools used by data scientists, analysts, machine learners, and software engineers. This is a free Python library that contains some of the algorithms that you will frequently use, such as classification, clustering, and regression algorithms. They include k-means clustering, support vector machines, random forests, gradient boosting and many more. Furthermore, this library is designed to be used with other libraries, such as the Python numerical library known as NumPy and the scientific library known as SciPy. In order to install the library, simply type *pip install scikit-learn*. Remember to replace pip with the

keyword for your chosen package manager, if you prefer a different one.

2. NumPy: This library goes hand in hand with Scikit-learn due to the support it offers for multi-dimensional arrays. This includes various algorithms that operate purely on these arrays which are used for data storage and certain matrix operations. The reason why we need this Python extension is because the programming language wasn't initially designed for numerical computing operations. Since complex mathematical operations are part of any data scientist's daily work, NumPy is a must-have.

3. SciPy: NumPy and SciPy are frequently used together because they fill each other's weak points. This open source library is used for technical and scientific computing processes. It adds modules to Python that are used for optimization, interpolation, linear algebra, image processing and much more. The reason why SciPy compliments NumPy is because it builds on the array object that is part of the NumPy library. Furthermore, it also expands on other libraries and tools such as Matplotlit and Pandas.

4. Pandas: This Python library was created mainly for manipulating data as well as analyzing it. Its most important features include data structures and operations that are needed to manipulate tables, as well as time series. Take note that Pandas is also free software and that it has nothing to do with pandas. The name of the library

refers to "panel data", which describes data sets that involve a set of observations over a number of recorded time periods.

5. Matplotlib: Sometimes you might want to sit back and visualize your data without looking at confusing numbers to form a conclusion. Sometimes you have to present your findings to those who are not technically inclined and do not understand raw numbers. In either case you would benefit from using plots to build a graphical representation of your findings. This is what Matplotlib is for. It allows you to build plots by using an array and then visualize the data, including interactively. In case you don't know what a plot is, it is a graph that is often used in statistical analysis in order to demonstrate a connection between certain variables.

6. Jupyter: Finally, we have Jupyter which will act as your data science IDE. It is often used by machine learners and data scientists because of how user-friendly it is and because it works with any programming language. However, the most important aspect is the fact that it allows you to visualize your data straight in the environment instead of opening new windows or panels and being forced to switch them. You write your code, your algorithms, and look at your data. In addition, Jupyter also provides you with the ability to directly communicate with other users and collaborate with them by sharing project files. We will discuss Jupyter in a bit

more detail soon.

These Python tools should be more than enough for a beginner data scientist. You may feel overwhelmed, especially when you consider there are hundreds of packages and libraries that you will eventually use. However, for now, you should stick to the tools that form the foundation of nearly every project.

Before moving on, let's discuss one more component which we mentioned only briefly, namely Jupyter. This tool forms your development environment and you will use it for every project. If you are used to the usual IDE's that are designed with programming and software development in mind, like Atom and Visual Studio, you will notice that Jupyter is quite different and you might even feel a bit lost. Therefore, let's take a moment to discuss using this environment.

Jupyter

All the concepts and techniques you will practice with the help of this book will be done using Jupyter. In order to install it, all you need to do is download it and use the install command with your chosen package manager. The setup itself is very easy to follow, so just perform every step as requested by the installer. If you encounter any errors, then you probably downloaded the wrong installer, so make sure you choose the correct one based on your operating system and its version. Once the installation is complete, fire up the program with the following command:

jupyter notebook

You will notice that the tool isn't opened inside an application window, but in your browser instead. The next step is to press the "New" button in order to tell Jupyter which version of Python you are using. Keep in mind that in this book the chosen version is Python 3. Next, you should see a blank window which is your environment and interface.

The first thing you will notice with this particular environment is that you aren't dealing with a typical text editor. Instead, the code is organized in cell blocks that are executed one cell at a time. This design gives you the advantage of testing and verifying your code one small block at a time. Furthermore, you will notice that you have an input where you type your code and an output where the result is displayed. Now, let's type something to see how it all works:

In: print ("I'm testing Jupyter!")

In order to execute this line of code you should click on the play button. You can find it beneath the Cell tab. The result of the code will be displayed inside the output cell, followed by a blank input cell where you can continue typing. You will also notice a small plus symbol which is used to create more cells.

Summary

In this chapter you learned what it takes to set up your work environment in order to get started. You learned step by step how to install Python and the various options you have. You explored the option of using a scientific distribution instead of a manual installation and you learned how to setup a virtual environment. Remember that if you are the type of person always worrying about messing up, especially when working with code and terminal instructions, then you should use a virtual environment and eliminate all of your fears and doubts with one swift blow.

Furthermore, you explored a number of Python libraries, packages, and extensions which make data science a much more approachable field. All of these tools are free to download and install and they offer you a great deal of benefits. Keep in mind that this book relies on some of these libraries for the practical implementations. For instance, Scikit-learn is used throughout the book because it contains all the algorithms we are going to use, as well as all of the datasets. This way all you need to do is type a command to import them, instead of manually downloading them and setting them up.

Chapter 2: Data Munging with Pandas

Now that you're set up with all the tools you need and with a clean environment, it's time to begin with the data science process known as data munging. Sometimes known as data wrangling, this step is one of the most important ones in the entire data science pipeline. The basic concept behind it is that you need to process a set of data in order to be able to use it together with another set of data, or to analyze it. Essentially, you will make enough changes to an original dataset in order to make it more useful for your specific goals. This is a pre-processing step.

In order to understand this idea better, imagine that you have a set of data on which you need to apply a classification algorithm. However, you realize that you can't perform this step just yet because the dataset is a combination of continuous and categorical variables. This means that you have to modify some of these variables to match the correct format. The challenge here is that you are dealing with raw data and you cannot analyze it just yet. First you need to clean the data with various data munging techniques and tools.

In most real world scenarios you will be dealing with a great deal of data that is raw and cannot be analyzed just yet, unlike the datasets used for study and practice. This is why you need to clean the data and it can actually take you a great deal of time. In fact, many data scientists spend more time preparing the data instead of coding or running various algorithms. So how do we prepare our data? One of the most popular methods includes using the Pandas library which is used for data analysis and manipulation, as mentioned earlier. The purpose of this library in this case is to allow you to analyze raw, real world data a lot faster.

Keep in mind that the purpose of data munging is to gather enough information in order to be able to detect the pattern within it. Furthermore, data needs to be accurate in order to be useful to a data scientist or analyst in order to cut down on the time and resources needed to come up with meaningful results. With that in mind, the first step you need to take involves acquiring the data. After all, without data you can't do anything. However, before you gain access to it you need to understand that all data items are different and are not created equally. You will often have issues recognizing authentic data with an identified source. The second step is about joining the data once it has been extracted from every source. At this stage, the data needs to be modified and then combined in order to proceed at a later time with the analysis. Finally, we have the last step that involves cleaning the data. This stage is the main one. You will need to modify the data in order to obtain a format you can use. You might also have to perform optional steps like correcting noisy data or bad data that can negatively influence the results.

As you can already see, this step is essential although tedious and time consuming. However, you sometimes cannot avoid it. You will have to make sure that you have relevant, up to data information that doesn't contain any null values in order to select only the data you are interested in for analysis. Fortunately, Python together with Pandas are some of the most powerful tools you can use to aid you in data munging.

The Process

As already mentioned, even if data science projects are unique, the workflow is generally the same. It all begins with the acquisition of data. You can gather data in many ways. You can extract it from a database, from images, from spreadsheets and any other digital source that holds information. This is your raw data, however. You cannot use it for a proper analysis because among all that data you also have missing information and corrupted data. You need to first bring order to chaos by using Python data structures to turn the raw information into an organized data set made out of properly formatted variables. This dataset is then processed with the help of various algorithms.

Next, you can examine your data in order to come up with an early observation that you will later test. You will obtain new variables by processing the ones you currently have and move up the data science pipeline with the various techniques such as graph analysis and reveal the most valuable variables. You will now be able to create the first data model, however, your testing phase will tell you that you will have to apply a number of corrections to your data and therefore return to the data munging processes to rework each step. Keep in mind that in most cases the output you expect will not reflect the real output that you will receive. Theory doesn't always lead you in the direction you expect and that is why you need to process a number of different scenario and test to see what works.

Importing Datasets

Before we can do anything we need the actual data. In order to import a dataset we are going to use Pandas to access tabular information from various databases or spreadsheets. Essentially, this tool will build a data structure in which each row of the tabular file will be indexed, while also separating the variables so that we can manipulate the data. With that being said, we will work in Jupyter and type the following command to import Pandas into our environment and access a CSV file:

In: import pandas as pd

iris_filename = 'datasets-ucl-iris.csv'

iris = pd.read_csv(iris_filename, sep=',', decimal='.', header=None, names= ['sepal_length', 'sepal_width', 'petal_length', 'petal_width', 'target'])

(Source: The Iris Dataset https://scikitlearn.org/stable/auto_examples/datasets/plot_iris_dataset.html retrieved in October 2019)

As you can see, the first step of this process is to import the tool we are going to use. Whether you are going to use Pandas or Scikit-learn, it is not enough to just have it installed on your system. You need to import it into your project in order to have access to its functions and features. Next, we created a new file and named it, while also defining the character that will act as a separator and a decimal. In this example the new file will contain an open source dataset that has been used to teach new data scientists and machine learners for years. The dataset is called Iris and it contains 50 samples of three different species of Iris flowers. We also mentioned that we don't want to define a header because it is not needed in our example. What we did so far was create a new data item named iris, which is in fact a data frame when we discuss it in the context of working with Pandas. In this case a data frame is actually the same as a Python list or dictionary, but with a set of added features. Next, we want to explore what the data item contains by typing:

In: iris.head()

This is a simple instruction without any parameters. By default if we don't specify we are going to access the first five rows from the file. If you want more, then you simply need to mention the number of rows you want to access by typing it as an argument between the parentheses of the function. Next, we want to read the names of the columns to see what kind of information they contain:

In: iris.columns

Out: Index(['sepal_length', 'sepal_width', 'petal_length',

'petal_width', 'target'], dtype='object')

As you can see in the output, what we have for now is an index of each column name. The structure of the output looks like a list. Now, let's obtain the target column:

In: Y = iris['target']

Y

This is what you should see as the output.

Out:

0   Iris-setosa

1   Iris -setosa

2   Iris -setosa

3   Iris -setosa

...

149   Iris-virginica

Name: target, dtype: object

The "Y" in this result is a series typical of Pandas. What you should know is that it is nearly identical to an array, however, it is only unidirectional. Furthermore, you will notice that the index class is the same as the one for a dictionary. Next, we are going to make a request to extract the list of columns by using the index:

In: X = iris[['sepal_length', 'sepal_width']]

Now we have the data frame, which is a matrix instead of a unidimensional series. The reason why it is a matrix is that we asked to extract several columns at once and therefore we essentially obtained an array that is structured in columns and rows. Now let's also obtain all of the dimensions.

In: print (X.shape)

Out: (150, 2)

In: print (Y.shape)

Out: (150)

The result is now a tuple and we can analyze the size of the array in either dimension. These are the bare bones basics of manipulating a new dataset and performing some basic exploration. Let's move on to the next step and preprocess the data so that we can actually use it.

Data Preprocessing

Now that you know how to load a dataset, let's explore the procedures you need to take in order to preprocess all the information within it. First, we are going to assume that we need to perform a certain action on a number of rows. In order to use any function, we first need to setup a mask. Take note that in this case a mask is in fact a collection of Boolean values that determine the selected line. The practical example will clear up this notion, so let's get to it:

In: mask_feature = iris['sepal_length'] > 7.0

In: mask_feature

0	False
1	False
...	
146	True
147	True
148	True
149	False

As you can see we have chosen only the lines which contain a sepal length value that is greater than seven. These observations are declared with a Boolean value. Next, we are going to apply a mask in order to modify the iris virginica target and create a new label for it:

In: mask_target = iris['target'] == 'Iris-virginica'

In: iris.loc[mask_target, 'target'] = 'MyLabel'

Wherever the old iris virginica label appeared it will now be replaced with "MyLabel" as the new label. Take note that for this operation we need to use the "loc" function in order to gain access to the data with the help of the indexes. Now let's take the next step and see the labels that are contained by the target column:

In: iris['target'].unique()

Out: array(['Iris-setosa', 'Iris-versicolor', 'New label'], dtype=object)

Now let's group all of the columns together:

In: grouped_targets_mean = iris.groupby(['target']).mean()

grouped_targets_mean

Out:

In: grouped_targets_var = iris.groupby(['target']).var()

grouped_targets_var

With this step we have grouped the columns together by using the groupby function. Take note that this is similar to the "group by" command that you have in SQL. In the next input line we have also applied the mean function which of course calculates the mean value. Keep in mind that we can apply this method either to a single column or multiple columns at the same time. Furthermore, we can use the variance, count, or sum functions in order to gain different values. Take note that the end result you obtain is also a Pandas data frame, which means you can connect all of these operations together. In this example we are grouping all of our data observations by labels in order to be able to analyze the difference between all the values inside the groups. But what if we also have time series to deal with?

In case you aren't familiar with time series, you should know that they imply the analysis of a collection of data entries that appear in an order. This order is determined chronologically. In essence, you are dealing with a group of points that are distributed in time with an equal space dividing each one of them. You will frequently encounter time series because they are used in many fields, usually regarding statistical analysis. For instance, when you have to work with weather data you will find time series regarding the forecasting or the detection of sunspots.

The next challenge, however, is dealing with data entries that contain noise. Keep in mind that while these training datasets have been documented and processed for years, they tend to be very clean and in fact they require very little preprocessing and cleaning. However, in the real world, you will frequently deal with noisy data. In that case, the first thing we can do is use a rolling function, which looks like this:

In: smooth_time_series = pd.rolling_mean(time_series, 5)

For this process we are applying the mean function once more. Keep in mind that you don't necessarily have to use the mean. You can also go with the median value instead. In addition, you will notice that we request to only access five samples. Next, we are going to use the apply function to perform a number of operations on our columns and rows. This is a function that can be used for multiple purposes, so let's start by determining how many non-zero items we have per line:

In: iris.apply (np.count_nonzero, axis=1).head()

Out: 0   5

1   5

2   5

3   5

4   5

dtype: int64

Finally, the applymap function is then used to perform operations on the elements themselves. Let's say that we need to obtain the value of the length for every single string representation inside every cell:

In: iris.applymap (lambda el:len(str(el))).head()

In order to obtain these values, the cells are casted to a string and then we can determine the length.

Now that you have an idea about using Pandas for data preprocessing, let's also discuss the topic of data selection with the help of the same tool.

Data Selection

What do we mean when we are referring to data selection? Let's assume that you have a dataset with an index column which you need to access in order to modify it and work with it. For the sake of this example, we are going to presume that the index starts from 100, like so:

n,val1,val2,val3

100,10,10,C

101,10,20,C

102,10,30,B

103,10,40,B

104,10,50,A

As you can see, the first row is row number 0 and its index value is 100. Once you import the file, you will see an index column as usual, however, there's the possibility of changing it or using it by mistake. Therefore, it would be a good idea to split the column from the rest of the data in order to avoid making any mistakes when you are running low on coffee. We are going to use Pandas to select the column and break it apart from the rest:

In: dataset = pd.read_csv('a_selection_example_1.csv',

index_col=0) dataset

That's it! Now you can manipulate the values as usual, whether you select the values by index, column, or line locations. For instance, you can access the fourth value from the fifth line which in our example has an index value equal to 105. Here's how the selection looks:

In: dataset['val4'][105]

You might be tempted to consider this to be a matrix, however, it isn't, so make sure not to make the confusion. In addition, you should always determine the column you want to access before you specify the row. This way you won't make any mistakes when looking to gain access to a certain cell's value.

Preparing your data and learning some surface information about it can greatly help you along the line, so make sure to always dedicate some time for data munging and preprocessing. In the next chapter, we are going to continue going deeper in order to gain further insight about the data exploration and data science pipeline as a whole.

Summary

In this chapter you learned the most elementary steps you need to take before starting to explore and analyze the data. You learned about data munging, which involves the preparation of data and how to work with a real dataset. You imported a real dataset and performed a number of basic operations on it with the purpose of preparing it for analysis. Now you know how to find out some basic information about your data, and therefore have a much easier time exploring it further. Keep in mind that while these steps may feel boring, they will make your job far easier down the line. In addition, you will gain much better results if you pre-process and explore your data as much as possible before you start implementing various learning algorithms and techniques.

Chapter 3: The Data Science Pipeline

So far we have imported a dataset and performed a few preprocessing operations with a series of Pandas function. The next step is to dive into the data science workflow. Preprocessing is a useful step, however, it is only the beginning, or the pre-beginning in fact.

In this section we are going to go through a series of somewhat challenging phases, so make sure to research and read the documentation of every tool we are going to use. We are going to discuss certain aspects about data science, such as creating new data features, performing dimensionality reduction, running performance tests and much more. All of this can be challenging and even overwhelming to a beginner, so make sure to read up on everything line by line and continue practicing the operations.

Exploring Data

The first phase within the data science workflow is performing the exploratory analysis. You need to gain a more detailed understanding of the data you are going to work with. This implies learning about the dataset's features, the shape of the elements, and using everything in order to form your hypothesis so that you can continue with the next steps. In this section we are going to continue working with the Iris dataset because it is very beginner-friendly and you are already somewhat familiar with it. So start by importing the dataset once more and create a new file like you did in the previous section. Once you've performed that step we can start exploring and go where no aspiring data scientist has gone before. Puns aside, we are going to use the describe function first in order to learn a bit more about our dataset:

In: iris.describe()

Now you should have access to various features such as deviation, minimum and maximum values, and so on. We need to take this basic data and analyze it at a deeper level, so let's start by exploring a graphical representation of it. For now we are going to use the simple "boxplot" function to create the plot, and not bother with Matplotlib just yet.

In: boxes = iris.boxplot(return_type='axes')

Keep in mind that you don't have to perform this step. Visualization is actually optional, unless you have to present your findings to someone who isn't very mathematically inclined but he or she can understand charts and plots much easier. For the same reason, as a beginner you should visualize your data so that you can place your focus elsewhere. Now, let's observe the relation between our features. For this step we are going to use a similarity matrix like this:

In: pd.crosstab(iris['petal_length'] > 3.758667, iris['petal_width'] > 1.198667)

You will notice that we are simply calculating the number of times the petal length appears when compared to the petal width. We are making this comparison with the crosstab function. If you look at the results you will see these features are well-related to each other. You can observe this even better by creating a scatterplot like so:

In: scatterplot = iris.plot(kind='scatter', x='petal_width', y='petal_length', s=64, c='blue', edgecolors='white')

The scatterplot makes it easy to see that the petal width is connected to the length. When exploring data this way you can also use a different kind of graph, namely a histogram. In our case we are going to use one in order to see a display of the distribution of the values.

In: distr = iris.petal_width.plot(kind='hist', alpha=0.5, bins=20)

In this case we have selected twenty bins. If you aren't familiar with histograms you should know that bins refer to a variable's intervals. This is calculated by determining the square root of the total count of observations. That value represents the total number of bins.

New Features

Unfortunately, we are rarely so lucky to discover a close connection between certain features like we did in our basic example. That's when we need to apply a series of transformations. Let's say that you are trying to determine the value of a house. All you know for certain is the size of each room. You can use this information to create a new feature that represents the construction's volume. This transformation needs to be applied because we cannot observe the volume, however, we can observe features like length, width and height and then use these features to calculate the volume. Here's how all of this can be applied with code:

In: import numpy as np

from sklearn import datasets

from sklearn.cross_validation import train_test_split

from sklearn.metrics import mean_squared_error

cali = datasets.california_housing.fetch_california_housing()

X = cali['data']

Y = cali['target']

X_train, X_test, Y_train, Y_test = train_test_split(X, Y, train_size=0.8)

(Source: California Housing dataset

https://scikitlearn.org/stable/modules/generated/sklearn.datasets.fetch_california_housing.html retrieved in October 2019)

This time we imported a new dataset called California housing, which contains a great deal of data on the Californian housing market. In this example we are going to implement a regressor together with a mean absolute error with a value of 1.1575. If the code is difficult to understand and you don't know what a regressor is, don't worry about it at this time, we will discuss this later. All you need to understand for now is the concept we're discussing.

In: from sklearn.neighbors import KNeighborsRegressor

regressor = KNeighborsRegressor()

regressor.fit(X_train, Y_train)

Y_est = regressor.predict(X_test)

print ("MAE=", mean_squared_error(Y_test, Y_est))

Out: MAE= 1.15752795578

Now we need to try and reduce the value of the mean absolute error by implementing Z scores. This way we can perform the regression comparison and feature normalization. This process is also known as Z normalization because it seeks to map all of the original features to the new features we created. Let's continue:

In: from sklearn.preprocessing import StandardScaler

scaler = StandardScaler()

X_train_scaled = scaler.fit_transform(X_train)

X_test_scaled = scaler.transform(X_test)

regressor = KNeighborsRegressor()

regressor.fit(X_train_scaled, Y_train)

Y_est = regressor.predict(X_test_scaled)

print ("MAE=", mean_squared_error(Y_test, Y_est))

Out: MAE= 0.432334179429

The value of the mean absolute error has now been reduced from the previous value of approximately 1.15 to nearly 0.4, which is quite a great result. There are other methods that can be employed in order to minimize this value, however, the transformations required would be too complicated to implement at this state. What you should gain from this example is the fact that basic transformations can be easily applied and they can make your exploratory analysis much easier to conduct.

Dimensionality Reduction

In the real world you will often work with datasets that contain tens of thousands of data items, if not hundreds of thousands. Such datasets tend to also contain a large number of features, which means there will be some of them that you will not need. Just because the information exists, that doesn't mean it's useful. In some cases features are simply irrelevant and they just contribute to the noise. Noise is one of the elements which reduce the accuracy of your analysis and anything you can do to reduce it translates to an accuracy boost. When noise is caused by irrelevant features, your best option is to use dimensionality reduction methods.

As the name suggests, dimensionality reduction is all about reducing useless features and cutting back on the time it takes to process your data. In this section we are going to discuss a couple of techniques and algorithms you can use to eliminate the features you don't need.

Covariance Matrix

As mentioned earlier, you need to compare all of your features, or collections of features, in order to determine whether a relationship exists between them. You don't want to eliminate useful features. The covariance matrix is one of the techniques you'll be using to achieve this.

Dimensionality reduction implies the detection of relevant features, as well as the removal of the rest. Once you detect the ones that don't offer much, you can eliminate them. To demonstrate this concept, we are going to once again import the Iris dataset. Remember that this dataset contains four features for each observation, therefore a correlation matrix will yield some useful results.

In: from sklearn import datasets

import numpy as np

iris = datasets.load_iris()

cov_data = np.corrcoef(iris.data.T)

print (iris.feature_names)

print (cov_data)

And this is how your output should look as a result.

['sepal length (cm)', 'sepal width (cm)', 'petal length (cm)',

'petal width (cm)']

[[1. -0.10936925 0.87175416 0.81795363]

[-0.10936925 1. -0.4205161 -0.35654409]

[0.87175416 -0.4205161 1. 0.9627571]

[0.81795363 -0.35654409 0.9627571 1.]]

With the covariance matrix in place, let's create a visual representation of our results in order to have an easier time drawing conclusions. As a beginner, you should always use visualization methods because they are so much easier to read than pure numbers. This time we are going to import Matplotlib to draw the plot for us:

In: import matplotlib.pyplot as plt

img = plt.matshow(cov_data, cmap=plt.cm.rainbow)

plt.colorbar(img, ticks=[-1, 0, 1], fraction=0.045)

for x in range(cov_data.shape[0]):

for y in range(cov_data.shape[1]):

plt.text(x, y, "%0.2f" % cov_data[x,y],

size=12, color='black', ha="center", va="center")

plt.show()

You will notice that this time we haven't used a scatterplot or a histogram. In fact, we created a heat map. Take note of the most important value, which is 1. Every feature covariance has been normalized to a value of one in order to help us see the powerful connection between a number of features. By analyzing the heat map, you will notice that feature one has a strong relation to feature three, as well as four. The third feature is also strongly connected to the fourth feature. Finally, we have feature two which seems to have no relation to any of the other features. It is completely independent. Now that you know which features are useful and which one is irrelevant, you can cut some of the useless information.

Principal Component Analysis

The next step is to use an algorithm like the principal component analysis in order to define smaller features from their parent features. The new ones will be linear, however. This means that the output's first factor will have most of the variance. The second vector will have the most of the left over variance, and so on. The information will be aggregated to a new set of vectors that are formed after employing a principal component analysis.

The implementation relies on the fact that the vectors contain the data that comes from the input, and everything else is just noise. All you need to do in this case is to decide the number of vectors to have. The decision is made based on the variance, however. Let's take a look at the practical approach to this algorithm:

In: from sklearn.decomposition import PCA

pca_2c = PCA(n_components=2)

X_pca_2c = pca_2c.fit_transform(iris.data)

```
X_pca_2c.shape
```

Out: (150, 2)

```
In: plt.scatter(X_pca_2c[:,0], X_pca_2c[:,1], c = iris.target,
alpha=0.8, s=60, marker='o', edgecolors='white')
plt.show()

pca_2c.explained_variance_ratio_.sum()
```

Out:

0.97763177502480336

(Adapted from: https://educationalresearchtechniques.com/2018/10/24/factor-analysis-in-python/)

After the implementation, you will see that we only have two features in our output. The principal component analysis object is represented by the "n_components" object and its value is equal to two, which translates to what we just discussed.

We aren't going to dig deeper into the principal component analysis algorithm because it would just add an additional layer of confusion. For now, as a beginner, you should understand the concept of dimensionality reduction and not the inner workings of the algorithm. However, if you wish, you are encouraged to explore further. For now, we are going to move on to discussing another dimensionality reduction technique, namely the latent factor analysis.

Latent Factor Analysis

This concept is similar to the principal component analysis. The main idea here is that a latent factor always exists somewhere. Take note that a latent factor is just a variable, which cannot be observed through direct methods. We can only assume that our features are affected by a latent variable. This type of variable contains a specific kind of noise known as an arbitrary waveform generator. With that being said, let's see how this methodology is used for dimensionality reduction.

In: from sklearn.decomposition import FactorAnalysis

fact_2c = FactorAnalysis(n_components=2)

X_factor = fact_2c.fit_transform(iris.data)

plt.scatter(X_factor[:,0], X_factor[:,1], c=iris.target,

alpha=0.8, s=60, marker='o', edgecolors='white')

plt.show()

(Adapted from: https://educationalresearchtechniques.com/2018/10/24/factor-analysis-in-python/)

The difference in this case is that we establish the covariance between the variables in the output.

Outlier Detection

This next stage is one of the most important ones. Determining outliers is important because if we have any kind of erroneous information in our dataset, or partially incomplete data, adapting any new data will be extremely problematic. In turn, this issue can lead to algorithms that process faulty data and create inaccurate results.

So what is an outlier in this case? When we detect that a data point deviated from other data points we can compare them and establish that it is in fact an outlier. Let's discuss several cases where we have a different outlier in order to gain a better understanding of how to detect them and treat them.

Mainly there are three situations and each one is handled differently. Firstly, we will presume that the outlier is an infrequent appearance in whatever dataset we are working with. In this scenario, the information is based on another set of data from which it was extracted. Here we have a data sample which contains an outlier that is flagged as one because of its assumed rarity. This type of outlier is dealt with through a basic removal process.

In the second example, we have an outlier that frequently manifests itself. In this case it appears frequently. Whenever you experience similar occurrences, there's a sizable chance of encountering an error that affects the data sample generation. The problem here is that the algorithm's priority isn't the generalization, instead it focuses on learning the non-focused distribution. The outlier has to be eliminated.

The third situation involves a data point which is easy to conclude that it is in fact an error. Datasets often contain faulty data entries and they can easily cause inconsistencies in your data whenever you modify or manipulate the value. All you need to do in this case is delete the value and instruct the model to presume that it is a random loss. Another option is calculating and using an average value instead of the erroneous one. This is a preferable solution, however, if you find it difficult to implement it then simply delete the outlier.

Knowing these scenarios will help you understand which one you are experiencing, thus allowing you to have an easier time detecting the outlier and removing it. The first phase is to determine every single outlier and locate it. You can use two techniques to do this, though they are similar. You can either examine all separated variables individually, or all of them at once. These techniques are called the univariate and multivariate analysis.

You already had a brief introduction to the univariate method because you already worked with it unknowingly. If you recall, you created a graphical representation using a boxplot earlier. When this technique is used, you can determine which variable is in fact an outlier because you will see them as extremes. For instance, let's say you are observing the data description. If this observation is beneath the 25% ration or greater than 75%, you are probably looking at an outlier which can easily be seen in a box plot. The same applies if you are making Z-score observations, except in this case you are looking for a value above three. This method is one of the most efficient at detecting outliers, however, not all of them can be exposed with just one technique. This way you will only identify the extreme variables and others will escape undetected. In this case you should also consider the multivariate option, however, for now it's enough to stick to the basics and eliminate the most obvious outliers.

Summary

In this chapter you learned how to prepare your data for analysis and how to explore it. Once you imported your dataset you need to go through a series of preparation techniques in order to increase your chances of reaching an accurate result. Data exploration is crucial to the preparation of data, as well as various techniques such as dimensionality reduction. Furthermore, you learned how to apply these dimensionality reduction techniques in practice by implementing a covariance matrix and a principal component analysis algorithm.

In addition, you also learned about the importance of finding outliers within your data. Depending on the type of outliers and the learning algorithms you are using, they can severely impact the training process and the accuracy of the final result. In this chapter you learned about situations where they occur and why they should be eliminated.

Chapter 4: Supervised Learning Algorithms

Now that you understand the fundamentals of data preparation and manipulation we can start with the main course.

The algorithms used in data science can be divided into several categories, mainly supervised learning, unsupervised learning, and to some degree semi-supervised learning.

As the name suggests, supervised learning is aided by human interaction as the data scientist is required to provide the input and output in order to obtain a result from the predictions that are performed during the training process. Once the training is complete, the algorithm will use what it learned to apply on new, but similar data.

The idea of supervised learning can be compared to the way humans learn. As a student you are guided by a professor with examples. You work through those examples with his or her help until you are finally able to work on your own.

In this chapter we are going to focus on this type of learning algorithm. However, take note that their purpose is divided based on the problems they need to solve. Mainly there are two distinct categories, namely regression and classification. In the case of regression problems, your target is a numeric value, while in classification it is a class or a label. To make things clearer, an example of a regression task is determining the average value of houses in a given city. A classification task, on the other hand, is supposed to take certain data like the petal and sepal length, and based on that information determine which is the species of a flower.

With that in mind, let's start the chapter by discussing regression algorithms and how to work with them.

Regression

In data science, many tasks are resolved with the help of regression techniques. However, regression can also be categorized in two different branches, which are linear regression and logistic regression. Each one of them is used to solve different problems, however both of them are a perfect choice for prediction analyses because of the high accuracy of the results.

The purpose of linear regression is to shape a prediction value out of a set of unrelated variables. This means that if you need to discover the relation between a number of variables you can apply a linear regression algorithm to do the work for you. However, this isn't its main use. Linear regression algorithms are used for regression tasks. Keep in mind that logistic regression is not used to solve regression problems like the name suggests.

Instead, it is used for classification tasks.

With that being said, we are going to start by implementing a linear regression algorithm on the Boston housing dataset, which is freely available and even included in the Scikit-learn library. This dataset contains 506 samples, with 13 features and a numerical type target. Unlike in our previous examples, this time the dataset will not be used as a whole. We are going to break it into two sections, a training and a testing set. There are no rules set in stone regarding the ratio of the split, however, it is generally accepted that it is best to keep the training set with a 70%-80% data distribution, and then save 20%-30% for the testing process.

Linear Regression

For this example we are going to use the Scikit-learn library because it contains the Boston dataset, therefore you don't have to download anything. Let's start the process by importing the dataset and splitting our data into two sets:

In: from sklearn.datasets import load_boston

boston = load_boston()

from sklearn.cross_validation import train_test_split

X_train, X_test, Y_train, Y_test = train_test_split(boston.data, boston.target, test_size=0.2, random_state=0)

Pay attention to the train_test_split function because it is used to split the dataset into a training and a testing set. Keep in mind that it is also part of the Scikit-learn library, another reason why we are using this amazing tool. Furthermore, you will notice that the data is not randomly categorized. We need to specify certain parameters. In this case, by declaring the size of the test set to be equal to a value of 0.2 we also automatically declare that the training set should be 80% of the total data. You don't have to declare both parameters. In addition, we also have a random state parameter. Its purpose is to generate random numbers during the split.

Now, let's add in the regressor in order to predict the target value of the testing set. In addition, we will also measure the resulting accuracy and make sure there isn't too much noise influencing the result in a negative way.

In: from sklearn.linear_model import LinearRegression

regression = LinearRegression()

regression.fit(X_train, Y_train)

Y_pred = regression.predict(X_test)

from sklearn.metrics import mean_absolute_error

print ("MAE", mean_absolute_error(Y_test, Y_pred))

Out: MAE 3.82230762633

That's it! The result isn't the best and improvements can certainly be made. However, the purpose of this demonstration is to get you started with linear regression. For now you should focus on correct implementation. As you progress, you will learn more about optimization and fine tuning. Furthermore, you will also see that these algorithms involve two important factors, namely processing speed and prediction accuracy. Your choice is determining the balance between the two.

Logistic Regression

We will not focus on logistic regression in this chapter, however, you should understand what makes it different from linear regression and what defines it. The most important aspect you should always keep in mind is that there is no regressor involved, therefore it is usually implemented to solve classification problems, specifically those of a binary nature. Binary classification refers to performing a classification task when you only have two classes, in other words Boolean labels. The main goal of the Booleans labels is to offer us a true or false result so that we can conclude whether a result has been predicted or not.

Logistic regression may not be quite as popular as its regression solving brother, however, it has started being applied more and more in the medical field nowadays. However, a more popular method of solving binary classification problems is with the use of a Naive Bayse classifier, which we are going to discuss in the next section.

Naive Bayes Classifier

As mentioned, a popular algorithm for classification tasks is the Naive Bayes classifier. Take note that this algorithm can also be used to solve multiclass classification problems as well.

This classifier is a fairly old one, before data science entered the mainstream and everyone talked about Python. However, this doesn't mean that it's obsolete and no longer in use. On the contrary, it is highly popular when performing the categorization of textual information. In plain English, this means that the algorithm can process a number of documents and then categorize them appropriately based on the content. For instance, the classifier can determine which emails belong in your spam box. In other cases, this algorithm is also used in combination with support vector machines (we will discuss this soon) in order to process massive amounts of data with a greater efficiency.

There are three types of classifiers and you will use each one of them under certain circumstances. The first one is known as the Gaussian Naive Bayes, which is an algorithm that automatically presumes which features have a normal distribution and how they are related to each class. The second one is the Multinomial classifier, which is the most popular one when working with model events. This is the algorithm that is used to classify documents based on how often a certain term is found within the content. Lastly we have the Bernoulli classifier, which describes the inputs with binary variables. This algorithm is used for the same purpose as the multinomial classifier.

For the purpose of this beginner's guide to data science, we are not going to apply all three of these algorithms, however, you are free to use the knowledge you gathered so far to use the right tools and put them in application. In this section you will go through the demonstration on how to implement the Gaussian algorithm. This classifier is used more frequently than the other, therefore we will make it a priority.

To demonstrate the application we are going to use the Iris dataset once again:

In: from sklearn import datasets

iris = datasets.load_iris()

from sklearn.cross_validation import train_test_split

X_train, X_test, Y_train, Y_test = train_test_split(iris.data, iris.target, test_size=0.2, random_state=0)

In: from sklearn.naive_bayes import GaussianNB

clf = GaussianNB()

clf.fit(X_train, Y_train)

Y_pred = clf.predict(X_test)

Just like before, we have divided the dataset into a training set and a testing set with an 80 by 20 division ratio. Then we imported the actual implementation of the algorithm from the Scikit-learn library in order to perform the classification task. The final step is acquiring the classification report so that we can see the data and form our own conclusions. Here's how a typical classification report looks:

In: from sklearn.metrics import classification_report

print (classification_report(Y_test, Y_pred))

Out:

	precision	recall	f1-score	support
0	1.00	1.00	1.00	12
1	0.93	1.00	0.96	11
2	1.00	0.83	0.91	7
avg / total	0.97	0.97	0.97	35

Take note of the four different metrics we have in the report. The first one is the precision metric which shows us the number of labels that are relevant. Next, we have the recall measure that gives us a value which represents the comparison between the relevant results and the labels. The third metric is the f1-score. This one is only important if we are dealing with a dataset that isn't well-balanced. The final metric we are interested in is the support measure that determines how many samples a certain class contains. In conclusion, the results we obtain after implementing the classifier are quite good. Now let's move to an even more powerful classifier that is used to solve complex problems.

K-Nearest Neighbors

This algorithm is one of the easiest ones to work with, however, it can solve some of the most challenging classification problems. The k-nearest neighbor algorithm can be used in various scenarios that require anything from compressing data to processing financial data. It is one of the most commonly used supervised machine learning algorithms and you should do your best to practice your implementation technique.

The basic idea behind the algorithm is the fact that you should explore relation between two different training observations. For instance, we will call them x and y, and if you have the input value of x, you can already predict the value of y. The way this works is by calculating the distance of a data point in relation to other data points. The k-nearest point is selected based on this distance and then it is assigned to a specific class.

To demonstrate how to implement this algorithm we are going to work with a much larger dataset than before, however, we will not use everything in it. Once again, we are going to rely on the Scikit-learn library in order to gain access to a dataset known as the MNIST handwritten digits dataset. This is in fact a database that holds roughly 70,000 images of handwritten digits which are distributed in a training set with 60,000 images and a test set with 10,000 images. However, as already mentioned, we are not going to use the entire dataset because that would take too long for this demonstration. Instead we will limit ourselves to 1000 samples. Let's get started:

In: from sklearn.utils import shuffle

from sklearn.datasets import

from sklearn.cross_validation import train_test_split

import pickle

mnist = pickle.load(open("mnist.pickle", "rb"))

mnist.data, mnist.target = shuffle(mnist.data, mnist.target)

As usual, we first import the dataset and the tools we need. However, you will notice one additional step here, namely object serialization. This means we converted an object to a different format so that it can be used later but also reverted back to its original version if needed. This process is referred to as pickling and that is why we have the seemingly out of place pickle module imported. This will allow us to communicate objects through a network if needed. Now, let's cut through the dataset until we have only 1000 samples:

mnist.data = mnist.data[:1000]

mnist.target = mnist.target[:1000]

X_train, X_test, y_train, y_test = train_test_split(mnist.data, mnist.target, test_size=0.8, random_state=0)

In: from sklearn.neighbors import KNeighborsClassifier

KNN: K=10, default measure of Euclidean distance

clf = KNeighborsClassifier(3)

clf.fit(X_train, y_train)

y_pred = clf.predict(X_test)

Now let's see the report with the accuracy metrics like earlier:

In: from sklearn.metrics import classification_report

print (classification_report(y_test, y_pred))

And here are the results:

Out:

	precision	recall	f1-score	support
0.0	0.68	0.90	0.78	79
1.0	0.66	1.00	0.79	95
2.0	0.83	0.50	0.62	76
3.0	0.59	0.64	0.61	85
4.0	0.65	0.56	0.60	75
5.0	0.76	0.55	0.64	80
6.0	0.89	0.69	0.77	70
7.0	0.76	0.83	0.79	76
8.0	0.91	0.56	0.69	77
9.0	0.61	0.75	0.67	87
avg / total	0.73	0.70	0.70	800

The results aren't the best, however, we have only implemented the "raw" algorithm without performing any kind of preparation operations that would clean and denoise the data. Fortunately the training speed was excellent even at this basic level. Remember, when working with supervised algorithms or any algorithms for that matter, you are always trading accuracy for processing speed or vice versa.

Support Vector Machines

The SVM is one of the most popular supervised learning algorithms due its capability of solving both regression as well as classification problems. In addition, it has the ability to identify outliers as well. This is one all-inclusive data science algorithm that you cannot miss. So what's so special about this algorithm?

First of all, support vector machines don't need much processing power in order to keep up with the prediction accuracy. We discussed several times how you are always dealing with a balancing act between accuracy and speed. This algorithm, however, is in a league of its own and you won't have to worry too much about sacrificing training speed for accuracy or the other way around. Furthermore, support vector machines can be used to eliminate some of the noise as well while performing the regression or classification tasks.

This type of algorithm has many real world applications and that is why it is important for you to understand its implementation. It is used in facial recognition software, text classification, handwriting recognition software and so on. The basic concept behind it, however, simply involves the distance between the nearest points where a hyperplane is selected from the margin between a number of support vectors. Take note that what is known as a hyperplane here is the object that divides the information space for the purpose of classification.

To put all of this theory in application we are going to rely on the Scikit-learn library once again. The algorithm will be implemented in such a way to demonstrate the accuracy of the prediction in the case of identifying real banknotes. We mentioned earlier that support vector machines are effective when it comes to image classification, therefore this algorithm is perfectly suited for our goals. What we need to solve in this example is a simple binary classification problem because we need to train the algorithm to determine whether the banknote is valid or not.

The bill will be described using several attributes. Keep in mind that unlike the other classification algorithms, a support vector machine determines its decision limit by defining the maximum distance between the data points which are nearest to the relevant classes. However, we aren't looking to limit the decision, we just want to find the best one. The nearest points in this best decision are what we refer to as support vectors. With that being said, let's import a new dataset and several tools:

import numpy as np

import pandas as pd

import matplotlib.pyplot as plt

dataset = pd.read_csv ("bank_note.csv")

As usual, the first step is learning more about the data we are working with. Let's learn how many rows and columns we have and then obtain the data from the first five rows only:

print (dataset.shape)

print (dataset.head())

Here's the result:

	Variance	Skewness	Curtosis	Entropy	Class
0	3.62160	8.6661	-2.8073	-0.44699	0
1	4.454590	8.1674	-2.4586	-1.46210	0
2	3.86600	-2.6383	1.9242	0.10645	0
3	3.45660	9.5228	-4.0112	-3.59440	0
4	0.32924	-4.4552	4.5718	-0.98880	0

(Source: Based on https://archive.ics.uci.edu/ml/datasets/banknote+authenticati on retrieved in October 2019)

Now we need to process this information in order to establish the training and testing sets. This means that we need to reduce the data to attributes and labels only:

x = dataset.drop ('Class', axis = 1)

y = dataset ['Class']

The purpose of this code is to store the column data as the x variable and then apply the drop function in order to avoid the class column so that we can store it inside a 'y' variable. By reducing the dataset to a collection of attributes and labels we can start defining the training and testing data sets. Split the data just like we did in all the earlier examples. Next, let's start implementing the algorithm.

We need Scikit-learn for this step because it contains the support vector machine algorithm and therefore we can easily access it without requiring outside sources.

from sklearn.svm import SVC

svc_classifier = SVC (kernel = 'linear')

svc_classifier.fit (x_train, y_train)

pred_y = svc.classifier.predict(x_test)

Finally, we need to check the accuracy of our implementation. For this step we are going to use a confusion matrix which will act as a table that displays the accuracy values of the classification's performance. You will see a number of true positives, true negatives, as well as false positives and false negatives. The accuracy value is then determined from these values. With that being said, let's take a look at the confusion matrix and then print the classification report:

from sklearn.metric import confusion_matrix

print (confusion_matrix (y_test, pred_y)

This is the output:

[[160 1]

[1   113]]

Accuracy Score: 0.99

Now let's see the familiar classification report:

from sklearn.metrics import classification_report

print (classification_report(y_test, y_pred))

And here are the results of the report:

	precision	recall	f1-score	support
0.0	0.99	0.99	0.99	161
1.0	0.99	0.99	0.99	114
avg / total	0.99	0.99	0.99	275

Based on all of these metrics, we can determine that we obtained a very high accuracy with our implementation of the support vector machines. A score of 0.99 is almost as good as it can get, however, there is always room for improvement.

Summary

In this chapter we have focused on supervised learning algorithms and techniques. This category of learning techniques contains some of the most beginner-friendly algorithms and methods of exploring and analyzing data. The purpose of this chapter is to offer you a practical introduction to topics such as linear regression, K-nearest neighbors, support vector machines and much more. Each concept is discussed in detail because they will serve as the foundation for any aspiring data scientist and machine learner. Just make sure to go over the code for each algorithm and learning technique and study the tools that we are using for the implementation.

Chapter 5: Decision Trees

Decision trees are built similarly to support vector machines, meaning they are a category of supervised machine learning algorithms that are capable of solving both regression and classification problems. They are powerful and used when working with a great deal of data.

It is important for you to learn beyond the barebones basics so that you can process large and complex datasets. Furthermore, decision trees are used in creating random forests, which is arguably the most powerful learning algorithm. In this chapter we are going to exclusively focus on decision trees explicitly because of their popular use and efficiency.

Implementing Decision Trees

Decision trees are essentially a tool which supports a decision that will influence all the other decisions that will be made. This means that everything from the predicted outcomes to consequences and resource usage will be influenced in some way. Take note that decision trees are usually represented in a graph, which can be described as some kind of chart where the training tests appear as a node. For instance, the node can be the toss of a coin which can have two different results. Furthermore, branches sprout in order to individually represent the results and they also have leaves which are in fact the class labels. Now you see why this algorithm is called a decision tree. The structure resembles an actual tree. As you probably guessed, random forests are exactly what they sound like. They are collections of decision trees, but enough about them.

Decision trees are one of the most powerful supervised learning methods you can use, especially as a beginner. Unlike other more complex algorithms they are fairly easy to implement and they have a lot to offer. A decision tree can perform any common data science task and the results you obtain at the end of the training process are highly accurate. With that in mind, let's analyze a few other advantages, as well as disadvantages, in order to gain a better understanding of their use and implementation.

Let's begin with the positives:

1. Decision trees are simple in design and therefore easy to implement even if you are a beginner without a formal education in data science or machine learning. The concept behind this algorithm can be summarized with a sort of a formula that follows a common type of programming statement: If this, then that, else that. Furthermore, the results you will obtain are very easy to interpret, especially due to the graphic representation.
2. The second advantage is that a decision tree is one of the most efficient methods in exploring and determining the

most important variables, as well as discovering the connection between then. In addition, you can build new features easily in order to gain better measurements and predictions. Don't forget that data exploration is one of the most important stages in working with data, especially when there's a large number of variables involved. You need to be able to detect the most valuable ones in order to avoid a time consuming process, and decision trees excel at this.

3. Another benefit of implementing decision trees is the fact that they are excellent at clearing up some of the outliers in your data. Don't forget that outliers are noise that reduces the accuracy of your predictions. In addition, decision trees aren't that strongly affected by noise. In many cases outliers have such a small impact on this algorithm that you can even choose to ignore them if you don't need to maximize the accuracy scores.

4. Finally, there's the fact that decision trees can work with both numerical as well as categorical variables. Remember that some of the algorithms we already discussed can only be used with one data type or the other. Decision trees, on the other hand, are proven to be versatile and handle a much more varied set of tasks.

As you can see, decision trees are powerful, versatile, and easy to implement, so why should we ever bother using anything else? As usual, nothing is perfect, so let's discuss the negative side of working with this type of algorithm:

1. One of the biggest issues encountered during a decision tree implementation is overfitting. Take note that this algorithm tends to sometimes create very complicated decision trees that will have issues generalizing data due to their own complexity. This is known as overfitting and it is encountered when implementing other learning algorithms as well, however, not to the same degree. Fortunately, this doesn't mean you should stay away from using decision trees. All you need to do is invest some time to implement certain parameter limitations to reduce the impact of overfitting.
2. Decision trees can have issues with continuous variables. When continuous numerical variables are involved, the decision trees lose a certain amount of information. This problem occurs when the variables are categorized. If you aren't familiar with these variables, a continuous variable can be a value that is set to be within a range of numbers. For example, if people between ages 18 and 26 are considered of student age, then this numerical range becomes a continuous variable because it can hold any value between the declared minimum and maximum.

While there are some disadvantages that can add to additional work in the implementation of decision trees, the advantages still outweigh them by far.

Classification and Regression Trees

We discussed earlier that decision trees are used for both regression tasks as well as classification tasks. However, this doesn't mean you implement the exact same decision trees in both cases. Decision trees need to be divided into classification and regression trees. They handle different problems, however, they are similar in some ways since they are both types of decision trees.

Take note that classification decision trees are implemented when there's a categorical dependent variable. On the other side, a regression tree is only implemented in the case of a continuous dependent variable. Furthermore, in the case of classification tree, the result from the training data is in fact the mode of the total relevant observations. This means that any observations that we cannot define will be predicted based on this value, which represents the observation which we identify most frequently.

Regression trees on the other hand work slightly differently. The value that results from the training stage is not the mode value, but the mean of the total observations. This way the unidentified observations are declared with the mean value which results from the known observations.

Both types of decision trees undergo a binary split however, going from the top to bottom. This means that the observations in one area will spawn two branches that are then divided inside the predictor space. This is also known as a greedy approach because the learning algorithm is seeking the most relevant variable in the split while ignoring the future splits that could lead to the development of an even more powerful and accurate decision tree.

As you can see, there are some differences as well as similarities between the two. However, what you should note from all of this is that the splitting is what has the most effect on the accuracy scores of the decision tree implementation. Decision tree nodes are divided into subnodes, no matter the type of tree. This tree split is performed in order to lead to a more uniform set of nodes.

Now that you understand the fundamentals behind decision trees, let's dig a bit deeper into the problem of overfitting.

The Overfitting Problem

You learned earlier that overfitting is one of the main problems when working with decision trees and sometimes it can have a severe impact on the results. Decision trees can lead to a 100% accuracy score for the training set if we do not impose any limits. However, the major downside here is that overfitting creeps in when the algorithm seeks to eliminate the training errors, but by doing so it actually increases the testing errors. This imbalance, despite the score, leads to terrible prediction accuracy in the end result. Why does this happen? In this case the decision trees grow many branches and that's the cause of overfitting. In order to solve this use, you need to impose limitations to how much the decision tree can develop and how many branches it can spawn. Furthermore, you can also prune the tree to keep it under control, much like how you would do with a real tree in order to make sure it produces plenty of fruit.

In order to limit the size of the decision tree, you need to determine new parameters during the definition of the tree. Let's analyze these parameters:

1. min_samples_split: The first thing you can do is change this parameter to specify how many observations a node will require in order to be able to perform the splitting. You can declare anything with a range of one sample to maximum samples. Just keep in mind that in order to limit the training model from determining the connections that are very common to a particular decision tree you need to increase the value. In other words, you can limit the decision tree with higher values.
2. min_samples_leaf: This is the parameter you need to tweak in order to determine how many observations are required by a node, or in other words a leaf. The overfitting control mechanism works the same way as for the samples split parameter.
3. max_features: Adjust this parameter in order to control the features that are selected randomly. These features are the ones that are used to perform the best split. In order to determine the most efficient value you should

calculate the square root of the total features. Just keep in mind that in this case, the higher value tends to lead to the overfitting problem we are trying to fix. Therefore, you should experiment with the value you set. Furthermore, not all cases are the same. Sometimes a higher value will work without resulting in overfitting.
4. max_depth: Finally, we have the depth parameter which consists of the depth value of the decision tree. In order to limit the overfitting problem, however, we are only interested in the maximum depth value. Take note that a high value translates to a high number of splits, therefore a high amount of information. By tweaking this value you will have control over how the training model learns the connections in a sample.

Modifying these parameters is only one aspect of gaining control of our decision trees in order to reduce overfitting and boost performance and accuracy. The next step after applying these limits is to prune the trees.

Pruning

This technique might sound too silly to be real, however, it is a legitimate machine learning concept that is used to improve your decision tree by nearly eliminating the overfitting issue. As with real trees, what pruning does is reduce the size of the trees in order to focus the resources on providing highly accurate results.

However, you should keep in mind that the segments that are pruned are not entirely randomly selected, which is a good thing. The sections that are eliminated are those that don't really help with the classification process and don't lead to any performance boosts. Less complex decision trees lead to a better optimized model.

In order to better understand the difference between an unmodified decision tree and one that was pruned and optimized, you should visualize the following scenario. Let's say that there's a highway that has a lane for vehicles that travel at 80 mph, and a second lane for the slower vehicles that travel at 50 mph. Now let's assume you are on this highway in a red car and you are facing a decision. You have the option to move on the fast lane in order to pass a slow moving car, however, this means that you will have a truck in front of you that can't achieve the high speed he should have in the left lane and therefore you will be stuck on that lane. In this case, the cars that are in the other lane are slowly starting to overpass you because the truck can't keep up. The other option is staying in your lane without attempting to make a pass. The most optimal choice here is the one that allows you to travel a longer distance during a certain amount of time. Therefore if you choose to stay in the slow lane until you gradually pass the truck that is blocking the fast lane, you will eventually be able to switch to that lane and pass all the other vehicles. As you can see, the second option might look slow at the time of consideration, however, in the long run it ends up being the most efficient one. Decision trees are the same. If you apply limits to your trees, they won't get greedy by switching you to the left lane where you will be stuck behind a truck. However, if you prune the decision tree, it will allow you to examine your surroundings in more detail and allow you predict a higher number of options you have in order to be able to make a better choice.

As you can see, performing the pruning process does yield a number of benefits which cannot be ignored. However, the implementation of this technique requires a number of steps and conditions. For instance, for a decision tree to be suitable for pruning, it needs to have a high depth value. Furthermore, the process needs to start at the bottom in order to avoid any negative returns. This issue needs to be avoided because if we have a negative node split at the bottom and another one occurs at the top, we will end up with a decision tree that will stop when the first division occurs. If the tree is pruned, it will not stop there and you will have higher gains.

Visualizing decision trees can sometimes be difficult when all you have is theory, so let's start with a step by step implementation to see them in action.

Decision Tree Implementation

Creating a decision tree starts from the root node. The first step is to select one of the data attributes and set up a logical test based on it. Once you have a set of results you can branch out and create another set of tests, which will you will use to create the subnode. Once we have at least a subnode we can apply a recursive splitting process on it in order to determine that we have clean decision tree leaves. Keep in mind that the level of purity is determined based on the number of cases that sprout from a single class. At this stage you can start pruning the tree in order to eliminate anything that doesn't improve the accuracy of the classification stage. Furthermore, you will also have to evaluate every single split that is performed based on each attribute. This step needs to be performed in order to determine which is the most optimal attribute, as well as split.

But enough theory for now. All you should focus on at this point is the fundamental idea behind decision trees and how to make them efficient. Once you think you grasped the basics, you need to start the implementation.

In the following example we will once again rely on the Iris dataset for the data, and the Scikit-learn library which contains it.

With that being said, let's continue with the practical implementation before the theory becomes too overwhelming. What's important at this stage is to understand the fundamental concepts and more importantly how to apply them in practice. With that being said, let's see what a decision tree looks like in code. Once again, we are going to use the Iris dataset, which is part of the Scikit-learn library. We will use other packages as well, such as Pandas and Numpy, so let's take a look at the code and discuss further:

```
%matplotlib inline

import pandas as pd

import numpy as np

import seaborn as sns

import matplotlib.pyplot as plt

from sklearn.model_selection import train_test_split

from sklearn.tree import DecisionTreeClassifier

df = pd.read_csv('Iris.csv')
```

Once all packages and modules are imported, we need to check the values to see if we have any null values:

```
df.isnull().any()
```

Take note that there are no null values in the dataset, however, this check should always be performed just to be certain. Always remember that data exploration is one of the most crucial steps and you should never avoid performing it. Make sure to perform all the other exploration steps we discussed earlier in the book in order to be as familiar with the data as possible. Once you consider you have analyzed and explored the data in enough detail, you can perform the analysis to determine the connection between the data columns. For this purpose in this example we are using the "seaborn" module because it contains the "pairplot" function, which will allow us to visualize all the connections. We will use a certain color for each column and then look for any outliers. Here's how to use this function:

sns.pairplot(df, hue = 'Species')

Take note that we do have a small number of outliers. In this case, they can probably be ignored because they are only a few and they most likely represent data anomalies and incorrect entries. For the purpose of this example we are going to assume they are nearly irrelevant anomalies and not pay them much attention.

The next step is performing the train / test split. Keep in mind that while in all other examples we have used a 80 / 20 ratio, this time we are going to have 30% of the data saved for the test set. Now let's take a look at the code:

all_inputs = df [['SepalLengthCm', 'SepalWidthCm', 'PetalLengthCm', 'PetalWidthCm']].values

all_classes = df ['Species'].values

(train_inputs, test_inputs, train_classes, test_classes) =

train_test_split (all_inputs, all_classes, train_size = 0.7, random_state = 1)

Notice that the random state is set to a value of one in order to make sure we always have an identical data split. This is not crucial for our decision tree example, however, if we ever want to recreate the dataset in the exact same way, we need this value.

Now, let's handle the classification process. We can now finally implement the decision tree:

dtc = DecisionTreeClassifier()

dtc.fit (train_inputs, train_classes)

dtc.score (test_inputs, test_classes)

The result should look something like this:

0.955

That's it! This is a simple example, but we already managed to achieve 95% accuracy. Now imagine if we would dedicate some time to polishing the decision tree as much as possible. This is why they are so frequently used to solve complex problems. Ultimately, they are easy to implement even if the theory makes it sound difficult, and they offer great results with minimal negative side effects.

Summary

In this chapter you explored decision trees and learned how they are implemented. You studied concepts such as backpropagation and pruning and you learned what advantages and disadvantages you will encounter when working with classification and regression trees. Remember that decision trees are powerful and very useful, however they sometimes need to be used together with other technique in order to enhance some of their characteristics. While they are easy to implement, you should study them in detail because they are often used in data science and Big Data, especially in the form of Random Forests. In this book we aren't going to expand on the concept of Random Forests, however, you should know that they are powerful and versatile in dealing with complex data because they are collections of decision trees. So, make sure you fully understand the concept of decision trees before progressing to more complex topics.

Chapter 6: Unsupervised Learning with K-means Clustering

Unlike supervised learning, this category of algorithms learns only from already established examples without human intervention. These algorithms are often used to process massive datasets that contain only inputs. Their objective is to identify the clusters of data points in order to compare them next to those coming from other datasets.

Unsupervised learning algorithms learn differently from supervised algorithms. They learn from unlabeled data, which hasn't been classified. They seek to define an output based on the common relations found in the inputs. That output is then applied to other datasets as well. Because of their learning mechanism, these algorithms are often used for tasks like explaining data features and statistical density estimations. Unsupervised learning can also be used anywhere where we need to discover what kind of anomalies the input data contains.

Unsupervised learning algorithms are mainly used to learn from complex datasets. Therefore, their top priority is performing an exploratory analysis and identifying important data. One such algorithm is known as the K-means Clustering algorithm and this chapter will focus on it alone. This learning technique is one of the most popular clustering algorithms that is used to discover predictable patterns in large datasets.

K-means Clustering

As mentioned, unsupervised learning methods are ideal for working with unlabeled data. However, to be more specific, one of the best techniques, if not the best, is to use a type of clustering algorithm. The main idea behind this approach is the cluster analysis which involves reducing data observations to clusters, or subdivisions of data, where each cluster contains information that is similar to that of a predefined attribute. Clustering involves a number of techniques that all achieve the same goal because they are all about forming a variety of theories regarding the data structure.

One of the most popular unsupervised learning algorithms and clustering techniques is known as k-means clustering. This concept revolves around building data clusters based on the similarity of the values. The first step is to determine the value of k, which is in fact represented by the total number of clusters we define. These clusters are built as k-many points, which hold the average value that represents the entire cluster. Furthermore, the values are assigned based on the value which is the closest average. Keep in mind that clusters have a core which is defined as an average value which pushes the other averages aside, changing them. After enough iterations, the core value will shift itself to a point where the performance metric is lower. When this stage is reached, we have the solutions because there aren't any observations available to be designated.

If you're confused by now by all this theory, that's ok. You will see that this technique is a lot easier than it sounds. Let's take a look at the practical implementation. In this example we will use the UCI handwritten digits dataset. It is freely available and you don't need to download if you are using Scikit-learn along with the book. With that being said, here's the code:

```
from time import time

import numpy as np

import matplotlib.pyplot as plt

from sklearn import datasets

np.random.seed()

digits = datasets.load_digits()

data = scale(digits.data)

n_samples, n_features = data.shape

n_digits = len(np.unique(digits.target))

labels = digits.target

sample_size = 300

print("n_digits: %d, \t n_samples %d, \t n_features %d"

% (n_digits, n_samples, n_features))
```

```
print(79 * '_')
print('% 9s' % 'init'    time  inertia  homo  compl  v-meas  ARI    AMI  silhouette')
def bench_k_means(estimator, name, data):
    t0 = time()
    estimator.fit(data)
    print('% 9s %.2fs %i %.3f %.3f %.3f %.3f %.3f %.3f'
          % (name, (time() - t0), estimator.inertia_,
             metrics.homogeneity_score(labels, estimator.labels_),
             metrics.completeness_score(labels, estimator.labels_),
             metrics.v_measure_score(labels, estimator.labels_),
             metrics.adjusted_rand_score(labels, estimator.labels_),
             metrics.silhouette_score(data, estimator.labels_,
                                      metric='euclidean',
                                      sample_size=sample_size)))
```

(Source: K Means clustering – Implementing k Means. Adapted from https://techwithtim.net/tutorials/machine-learning-python/k-means-2/ accessed in October 2019)

If you analyze the code line by line you will notice that the implementation is fairly simple, logical, and easy to understand. In fact, it is somewhat similar in parts to other techniques we used so far. However, there is one major difference that is important to mention, namely the performance measurements we are using in order to accurately interpret the data.

First we have a homogeneity score. This metric can have a value between zero and one. It mainly seeks the clusters that make room only for one class systems. The idea is that if we have a score that is close to the value of one, then the cluster is mostly built from the samples that belong to a single class. On the other hand, if the score is close to zero, then we have achieved a low homogeneity.

Next, we have the completeness score. This metric compliments the homogeneity measure. Its purpose is to give us information on how the measurements became part of a specific class. The two scores allow us to form the conclusion that we either managed to perform perfect clustering, or we simply failed.

The third metric is known as the V-metric, or sometimes the V-measure. This score is calculated as the harmonic mean of the previous two scores. The V-metric essentially checks on the homogeneity and the completeness score by assigning a zero to one value that verifies the validity.

Next, we have the adjusted Rand index metric. This is a score that is used to verify the similarity of the labeling. Using a value between zero and one, the Rand index simply determines the relation between the distribution sets.

Finally, we have the silhouette metric which is used to verify whether the performance of the clustering is sufficient without having labeled data. The measurement goes from a value of negative one to positive one and it determines whether the clusters are well-structured or not. If the number is anywhere close to negative one then we are dealing with bad clusters. In order to make sure we have dense clusters, we need to achieve a score close to positive one. Keep in mind that in this case we can also have a score that is close to zero. In this case the silhouette measurement tells us that we have clusters which are overlapping each other.

Now that you understand the measurement system, we have to apply one more step to this implementation and make sure that the results are accurate. To verify the clustering scores we can use the bench_k_means function like so:

bench_k_means(KMeans(init='k-means++', n_clusters=n_digits, n_ init=10),

name="k-means++", data=data) print(79 * '_')

Now let's see what conclusion we can draw from the scores. Here's how your results should look:

n_digits: 10, n_samples 1797, n_features 64

init	time	inertia	homo	compl
k-means++	0.25s	69517	0.596	0.643

init	v-meas	ARI	AMI	silhouette
k-means++	0.619	0.465	0.592	0.123

As you can see, with a basic k-means implementation we have fairly decent scores, however, there is a lot we could improve. Clustering is sufficient, but we could perfect the scores by implementing other supervised or unsupervised learning techniques. For instance, in this case you might consider using the principal component analysis algorithm as well. Another option would be applying various dimensionality reduction methods. However, for the purpose of learning how to implement the K-means clustering algorithm, these results will suffice. However, you should keep in mind that in the real world of data science you will often implement a number of algorithms and techniques together. You will almost never be able to get useful results with just one algorithm, especially when working with raw datasets instead of the practice ones.

Summary

In this chapter we have focused on learning about unsupervised learning algorithms, namely K-means clustering. The purpose of this section was to show you a technique that can be used on more complex datasets. Clustering algorithms are the staple of data science and frequently used, especially in combination with other algorithms and learning techniques. Furthermore, as you will learn later, clustering techniques, especially K-means clustering, are highly efficient in dealing with Big Data.

Chapter 7: Neural Networks

Artificial neural networks were designed theoretically in the 1950s, with the purpose of finding a way to copy the human thinking and decision making process. The idea was to create a learning system that would function similarly to the human brain. However, keep in mind that this didn't involve the attempt to replicate the human brain or neural network. What the artificial and the natural systems had in common is only the shallow resemblance. Artificial neural networks were created using the inspiration that came from exploring the human neural network.

Neural networks consist of neurons, or nodes to be more precise. These nodes are in fact just code which seeks to find the solution to a problem through repeated execution. Once an answer is found, it is communicated to the next node which in turn will process that information. These nodes are setup in a system of layers. In other words, there are separate layers that consist of a certain number of nodes and these layers are separate from each other, though they do communicate. First the communication is done only through the nodes of the same layer and only then the information is passed on to the following layer. Based on this simple description, you can already determine that the artificial neural network is nowhere near the human neural net when it comes to capability and processing power. Data travels in a somewhat linear fashion, from node to node and layer to layer. Human processing however is parallel, which means that the data is processed at the same time throughout the vast interconnected network of neurons.

Now that you understand the basic concept behind a neural network we can dive into this chapter and discuss its design and the various types of neural networks that we can implement.

Neural Network Structures

So far you learned that when in a neural network data is explored, classified and then pushed to the next node, ultimately traveling through the whole network. We also mentioned layers, however, take note that there are three different categories of layers. First we have the input layer and the output layer. Between these layers however, we have what is known as a hidden layer. This is where the input layer sends all of the data it analyzes. Keep in mind that neural networks can also function without the hidden layer and only use the two basic input and output layers. On the other end of the spectrum we can also develop networks with hundreds of layers. In this chapter we are going to focus on the more classic design which is based on having the three layers. Generally speaking, if a neural network consists of more than these layers it turns into a deep learning system. This type of network is different and as a beginner there's no need to get into it just yet.

As you may have guessed already, the structure of neural networks is influenced by the way the nodes are arranged inside the network. With each arrangement we end up with a different neural network. The most common arrangements lead to the development of the feedforward neural networks, recurrent neural networks and the Boltzmann machine, just to list a few. Each one of these networks has a different structure and serves a different purpose.

Now let's discuss the structure of a basic network in more detail.

As already discussed, a common network has three layers, but not always. For instance, without any hidden layers, we start working with logistic regression. The input layer is designed to contain an observation matrix that has five units. The output layer, however, has three units. In the case of the output, we need to determine the difference between the three units, or classes, and therefore we have to classify them. In the case of the input, we are dealing with constrained numerical features which need to be turned into categorical features. The hidden layer on the other hand starts with eight units, but this is only the minimum value. There is no actual limitation to how many units there can be, and as already mentioned, there is no limit to the amount of hidden layers we have either. Due the lack of restrictions in this case, the data scientist needs to use his best judgment to make the right decision regarding the type of network he or she implements.

With the fundamentals fresh in your mind, let's discuss the first neural network model, which is the feedforward neural net.

The Feedforward Neural Network

This is the fundamental design of the neural network as it consists of the three basic elements that communicate in a line. The input layer sends data to the hidden layer, which in turn sends it to the output layer. Take note that this type of neural network does not seek to iterate. In other words, the data inside the input layer is processed and the result can be found only in the output layer. In addition, there are two categories of feedforward networks, namely the single layer perceptron and the multilayer perceptron.

The simplest type of neural network you can have is the single layer perceptron. This is a type of feedforward neural network that contains only one node per layer. The node has an input that is directly connected to the node of the next layer because there are no other nodes around it. The second node, however, is in fact the weighted sum of the inputs. The input data is defined by the node and if a certain condition is met, then we gain a result. The single layer perceptron is used only to classify the data into two different parts, therefore it is also known as a linear binary classifier.

Another simplistic neural network is the multilayer perceptron. In this case we have to have at least two layers, however, in a real world implementation we would have quite a few more. The main difference between this model and the single layer perceptron, however, is the fact that the output of one layer turns into the input of the following layer. This architecture is based on a concept called backpropagation (more on this soon). With this design we have to perform a comparison check between the output we assume and the real output. In the event that we find a high inaccuracy level, the data is transmitted once again through the entire network. During the second round of processing the node's weights will be adjusted in order to improve the accuracy level. Keep in mind that this process repeats itself multiple times until the result is accurate. While the single layer perceptron has limited use, as the first type of neural network ever devised, the multilayer perceptron has a wide range of usability because of this efficient structure.

Next up we have the recurrent neural network structures, however before moving on to that topic we should clear up what the concept of backpropagation involves.

Backpropagation

Backpropagation is a concept that appeared in the 70s when neural networks were mostly theoretical. Back then computers were simply not powerful and efficient enough to handle an artificial neural network, therefore, even though this technological advancement was already designed, it could not be applied. Even today working with a neural network requires a great deal of resources from your computer. But let's leave history in the past and see what backpropagation is all about.

This concept is in fact a mathematical concept and therefore we will not dig deep into the subject because we are only interested in the theory and what it means to you as an aspiring data scientist.

Remember that there is always a difference between the imagined output and the actual output. Furthermore, you should focus on the fact that each node has a weight assigned to it, which is adjusted based on the information that arrives from the input. The purpose of backpropagation is to find the off-shoot of this mathematical difference between the two outputs. This result is then applied to the next iteration in order to give the nodes the ability to adjust their weights. This is why every single time the data is processed through the entire network, the inaccuracy level is reduced.

Just in case you are having difficulty with this theory, you should imagine the following scenario. Let's say that you are out playing basketball. You throw the ball to score, but unfortunately it just hits the hoop itself and by doing so the ball just starts going around the ring. With every single pass, the ball will slow down and fall slightly inwards towards the center of the hoop. Eventually, after enough passes the ball will go down through the ring. This is how backpropagation can also be visualized. Every loop the ball completes is in fact an iteration. The more iterations we have, the closer we get to the result, or in our case, the closer the ball gets to the center of the ring. With each pass, the output becomes closer to what we expect.

In data science one of the main goals is to process data automatically with supervised and unsupervised machine learning methods. This same goal applied to artificial neural networks. There is one notable difference, however. The learning algorithms we discussed earlier require already documented information in order to be able to gain experience and learn on their own so that they can make future decisions. Neural networks, on the other hand, are capable of making decisions in the present, which makes them far more powerful and efficient. That is why neural networks are used when dealing with complex data and complicated patterns that require a system strong enough to lead to accurate results.

Recurrent Neural Networks

Now that you have a basic understanding of the simple feedforward neural networks and the concept of backpropagation, we can advance to recurrent neural networks. As you already know, the feedforward network's system works by sending data along a straight line, from node to node and layer to layer. The input is point A and the output is point B. On the other end we have the recurrent neural networks that transmit data both ways. While this is already an important change in the design of the network, it is not what makes it so special.

This type of network has the particular ability to remember all of the states a set of data goes through. After every iteration, the old state of the data will be memorized. Keep in mind that the original neural network design wasn't capable of memorization, meaning that with each iteration the previous data state would be lost. Imagine watching a video and after an ad pops up you forget what you already watched.

The structure of a recurrent neural network can be described as a series of linked nodes that consists of inputs and outputs. This system where an input becomes the output allows for highly accurate results. Now let's discuss one particular recurrent neural network that has been popularized by machine learners and data scientists in the last few years. In the next section we are going to explore the Boltzmann machine and see how to implement it in a real scenario.

The Restricted Boltzmann Machine

This neural network is a special representation of the recurrent network architecture because of the fact that it borrows certain design elements from the multilayer perceptron. For instance, the inputs and the hidden layers structures are the same as those of the perceptron, however, they are also interconnected. Take note that the structure doesn't force the network to communicate data in a particular way, namely linearly, because all elements are connected to each other to a certain degree. This means that the nodes can communicate with each other during every processing cycle, and therefore all the input variables are analyzed and modeled. Other networks only focus on the variables which can be observed. The main benefit of this system is the fact that it makes it possible for the output nodes and layer to advance.

The restricted Boltzmann machine neural network is a model that relies on an energy function. This function is what actually supports the relationship between each value and node configuration. Keep in mind that this energy value is in fact a metric which is determined mathematically and it is used to perform parameter updates. This also means that a free energy value will be present and one of your tasks will be to minimize it in order to create a powerful model.

The reason why we are discussing a restricted Boltzmann machine instead of the fundamental Boltzmann machine is because we don't want to face any scaling problems. With the simple architecture we would have a large number of nodes which would require a lot of computation power and time in order to process the data and reach a satisfactory output. This might not be a problem if we have an adequate amount of resources, however, with the increase in the number of nodes, the value of the free energy would also be higher. This value could become unmanageable, therefore, it is easier to go with a restricted model from the beginning, which implies a change in the structure, as well as the training stage. In order to restrict the network, all we need to do is apply a connectivity limit to all of the network's nodes. In other words, we need to make sure that we restrict the relation between all of the nodes that belong to a certain layer. In addition, we also need to remove the communication line between all the layers that are next to each other. These two limitations are what give this neural network its name.

Keep in mind that these restrictions aren't always the most optimal solution that will solve all of your scaling issues. Another problem we are facing is the time it takes to perform the model training. That is why we also need to implement a machine learning technique known as the Permanent Contrastive Divergence in order to lower the value of the free energy even more.

Now that the theory is out of the way, let's see this popular recurrent neural net in action in the next section and discuss the implementation of the code.

Restricted Boltzmann Machine in Practice

As mentioned, this is one of the most powerful models you can implement, especially as a beginner. You should only use this type of neural network when working with complex datasets that contain a great deal of information. To demonstrate the application of the restricted Boltzmann machine, we are going to use the MNIST handwritten digits dataset. The model will be implemented by first defining an object and its layers, vectors and various functions and attributes that are used to communicate data between the layers. Furthermore, we will apply the contrastive divergence solution in order to improve the time it takes to perform the training process.

```
class RBM(object):

    def __init__(
        self,
        input=None,
        n_visible=784,
        n_hidden=500,
        w=None,
        hbias=None,
        vbias=None,
        numpy_rng=None,
        theano_rng=None
    ):
```

Let's take a break and discuss what we did so far. We have defined a new object that stands for the Boltzmann machine. As you can see there are a number of parameters attributed to this object. They define how many network nodes we have, including those that are visible and invisible, inside the hidden layer. Furthermore we have a few other attributes that are not obligatory, however, they will help in creating a better model.

self.n_visible = n_visible

self.n_hidden = n_hidden

if numpy_rng is None:

    numpy_rng = numpy.random.RandomState(1234)

if theano_rng is None:

    theano_rng = RandomStreams(numpy_rng.randint(2 ** 30))

if W is None:

initial_W = numpy.asarray(

    numpy_rng.uniform(

        low=-4 * numpy.sqrt(6. / (n_hidden + n_visible)),

        high=4 * numpy.sqrt(6. / (n_hidden + n_visible)),

        size=(n_visible, n_hidden)

),

dtype=theano.config.floatX

)

The "w" attribute is used here so that we can enable the use of the GPU. In this example the GPU can handle a great deal more data processing than the CPU. It is one of the simple solutions we can use to speed up the training process. In addition we also allow data sharing between various functions that contain "theano.shared". The next step is to allow this sharing process throughout the network.

W = theano.shared(value=initial_W, name='W', borrow=True)

if hbias is None:

hbias = theano.shared(

value=numpy.zeros(n_hidden,dtype=theano.config.floatX

),

name='hbias',

borrow=True

)

if vbias is None:

```python
vbias = theano.shared(
    value=numpy.zeros(n_visible,
    dtype=theano.config.floatX
    ),
    name='vbias',
    borrow=True
)
```

Next, let's setup the input layer.

```python
self.input = input

if not input:

    self.input = T.matrix('input')

self.W = W

self.hbias = hbias

self.vbias = vbias

self.theano_rng = theano_rng

self.params = [self.W, self.hbias, self.vbias]

def propup(self, vis):
```

```python
pre_sigmoid_activation = T.dot(vis, self.W) + self.hbias

return [pre_sigmoid_activation, T.nnet.sigmoid(pre_sigmoid_activation)]

def propdown(self, hid):

pre_sigmoid_activation = T.dot(hid, self.W.T) + self.vbias

return [pre_sigmoid_activation, T.nnet.sigmoid(pre_sigmoid_activation)]
```

This part of the code also includes the function which communicates the activation of visible units in an upward direction towards the hidden units. This way the hidden units can determine their activation based on the samples from the visible units. This process is also performed downwards. Next, we are going to transmit the hidden layer's activations as well as the visible layer's activations.

```python
def sample_h_given_v(self, v0_sample):

pre_sigmoid_h1, h1_mean = self.propup(v0_sample)

h1_sample = self.theano_rng.binomial(size=h1_mean.shape,

n=1, p=h1_mean, dtype=theano.config.floatX)

return [pre_sigmoid_h1, h1_mean, h1_sample]

def sample_v_given_h(self, h0_sample):
```

```
pre_sigmoid_v1, v1_mean = self.propdown(h0_sample)

v1_sample = self.theano_rng.binomial(size=v1_mean.shape,

n=1, p=v1_mean, dtype=theano.config.floatX)

return [pre_sigmoid_v1, v1_mean, v1_sample]
```

We have established the connections and therefore we can handle the sampling process.

```
def gibbs_hvh(self, h0_sample):

    pre_sigmoid_v1, v1_mean, v1_sample
    =self.sample_v_given_h(h0_sample)

    pre_sigmoid_h1, h1_mean, h1_sample
    =self.sample_h_given_v(v1_sample)

    return [pre_sigmoid_v1, v1_mean, v1_sample
            pre_sigmoid_h1, h1_mean, h1_sample]

def gibbs_vhv(self, v0_sample):

        pre_sigmoid_h1, h1_mean, h1_sample =
        self.sample_h_given_v(v0_sample)

        pre_sigmoid_v1, v1_mean, v1_sample =
        self.sample_v_given_h(h1_sample)

        return [pre_sigmoid_h1, h1_mean, h1_sample,
```

pre_sigmoid_v1, v1_mean, v1_sample]

As you can see, the implementation of a restricted Boltzmann machine is slightly more complicated than any other models, however, it is also far more powerful. So far we have established all of our layers and neural nodes and we have finished setting up the connection lines. Furthermore, we have taken care of the sampling process. Keep in mind that sampling is used for the implementation of the permanent contrastive divergent concept, which also requires a count parameter. In this phase of the implementation, handling the free energy value is top priority. Once all of this is achieved, all we need to do is check how successful our model is.

def free_energy(self, v_sample):

wx_b = T.dot(v_sample, self.W) + self.hbias

vbias_term = T.dot(v_sample, self.vbias)

hidden_term = T.sum(T.log(1 + T.exp(wx_b)), axis=1)

return -hidden_term - vbias_term

def get_cost_updates(self, lr=0.1, persistent = , k=1):

pre_sigmoid_ph, ph_mean, ph_sample = self.sample_h_given_v(self.input)

chain_start = persistent

```
(
    [
        pre_sigmoid_nvs,
        nv_means,
        nv_samples,
        pre_sigmoid_nhs,
        nh_means,
        nh_samples
    ],
    updates
) = theano.scan(
    self.gibbs_hvh,
    outputs_info=[None, None, None, None, None, chain_start],
    n_steps=k
)
chain_end = nv_samples[-1]
cost       =       T.mean(self.free_energy(self.input))       -
```

```
T.mean(self.free_energy(chain_end))

gparams = T.grad(cost, self.params,
consider_constant=[chain_end])

for gparam, param in zip(gparams, self.params):

updates[param] = param - gparam * T.cast(lr,
dtype=theano.config.floatX)

updates = nh_samples[-1]

monitoring_cost = self.get_pseudo_likelihood_cost(updates)

return monitoring_cost, updates
```

In this section of the code we have defined the permanent contrastive divergence algorithm and we have established the learning rate parameter as well as the k parameter. As it suggests, the learning parameter refers to the speed it takes for the algorithm to learn and it gives us the option to adjust it as needed. The k parameter on the other hand represents the amount of steps which the algorithm has to perform in order to determine the value of the free energy. Now let's write the code to establish the training process.

```
def get_pseudo_likelihood_cost(self, updates):

bit_i_idx = theano.shared(value=0, name='bit_i_idx') xi =
T.round(self.input)
```

```python
fe_xi = self.free_energy(xi)

xi_flip = T.set_subtensor(xi[:, bit_i_idx], 1 - xi[:, bit_i_idx])

fe_xi_flip = self.free_energy(xi_flip)

cost = T.mean(self.n_visible * T.log(T.nnet.sigmoid(fe_xi_flip - fe_xi)))

    updates[bit_i_idx] = (bit_i_idx + 1) % self.n_visible

return cost

train_rbm = theano.function(

[index],

cost,

updates=updates,

givens={x: train_set_x[index * batch_size: (index + 1) *batch_size] },

name='train_rbm'

)

plotting_time = 0.

start_time = time.clock()
```

(Source: Adapted from RBM tutorial code

http://deeplearning.net/tutorial/code/deep.py retrieved in October 2019)

That's it! We have successfully implemented a restricted Boltzmann machine. This neural network might be somewhat out of your skill range as a beginner, however, it offers you a glimpse of what's to come. Furthermore, you will see that if you analyze the code line by line and read and reread the theory you will understand that a recurrent neural network such as this isn't that difficult to setup. You can even continue exploring this topic on your own, as our model can be improved and extended. The results we obtained are good, but they can always be better.

Summary

In this chapter you learned about the main types of neural networks and where they fit in the world of data science. The purpose of this section was to educate you on topics such as perceptron and recurrent neural networks. While we mostly focused on beginner-friendly theory so that you can fully understand neural networks, we have also worked with an example of the Restricted Boltzmann Machine. Keep in mind that this topic is somewhat more complex and towards the intermediary level, however, you should understand how neural networks work because they are one of the main structures used to create advanced learning and training models. Just make sure to analyze each line of code because the implementation of the

Boltzmann machine can be quite lengthy when compared to basic supervised or unsupervised learning algorithms.

Chapter 8: Big Data

In data science, the purpose of supervised and unsupervised learning algorithms is to provide us with the ability to learn from complicated collections of data. The problem is that the data that is being gathered over the past few years has become massive in size. The integration of technology in every aspect of human life and the use of machine learning algorithms to learn from that data in all industries has led to an exponential increase in data gathering. These vast collections of data are known in data science as Big Data. What's the difference between regular datasets and Big Data? The learning algorithms that have been developed over the decades are often affected by the size and complexity of the data they have to process and learn from. Keep in mind that this type of data no longer measures in gigabytes, but sometimes in petabytes, which is an inconceivable number to some as we're talking about values higher than 1000 terabytes when the common household hard drive holds 1 terabyte of information, or even less.

Keep in mind that the concept of Big Data is not new. In fact, it has been theorized over the past decades as data scientists noticed an upward trend in the development of computer processing power, which is correlated with the amount of data that circulates. In the 70s and 80s when many learning algorithms and neural networks were developed, there were no massive amounts of data to process because the technology back then couldn't handle it. Even today, some of the techniques we discussed will not suffice when processing big data. That is why in this chapter we are going to discuss the growing issue of Big Data in order to understand the future challenges you will face as a data scientist.

The Challenge

Nowadays, the problem of Big Data has grown so much that it has become a specialized subfield of data science. While the previous explanation of Big Data was rudimentary with the purpose of demonstrating the problem we will face, you should know that any data is considered Big Data as long as it is a collection of information that contains a large variety of data that continues to grow at an exponential rate. This means that the data volume grows as such a speed that we can no longer keep up with it in order to process and analyze it.

The issue of Big Data appeared before the age of the Internet and online data gathering, and even if today's computers are so much more powerful than in the pre-Internet era, data is still overwhelming to analyze. Just look around you and really focus on how many aspects of your life involve technology. If you stop to think you will realize that even objects that you never considered as data recorders, save some kind of data. Now this thought might make you paranoid, however, keep in mind that most technology records information regarding its use and the user's habits in order to find better ways to improve the technology. The big problem here is that all of this technology generates too much data at a rapid pace. In addition, think about all the smart tech that's being implemented into everyone's homes in the past years. Think of Amazon's Alexa, "smart" thermostats, smart doorbells, smart everything. All of this records data and transmits it. Because of this, many professional data scientists are saying that the current volume of data is being doubled every two years. However, that's not all. The speed at which this amount of data is generated is also increasing roughly every two years. Big Data can barely be comprehended by most tech users. Just think of the difference between your Internet connection today and just five year ago. Even smartphone connections are now powerful and as easy to use as computers.

Keep in mind that we are dealing with a vicious circle when it comes to Big Data. Larger amounts of data generated at higher speeds mean that new computer hardware and software has to be developed in order to handle the data. The development of computer processing power needs to match the data generation. Essentially, we are dealing with a complex game of cat and mouse. To give you a rough idea about this issue, imagine that back in the mid-80s the entire planet's infrastructure could "only" handle around 290 petabytes of information. In the past 2 years, the world has reached a stage where it generates almost 300 petabytes in a span of 24 hours.

What all that in mind, you should understand that not all data is the same. As you probably already know, information comes in various formats. Everything generates some kind of data. Think of emails, crypto currency exchanges, the stock market, computer games, and websites. All of these create data that needs to be gathered and stored, and all of it ends up in different formats. This means that all of the information needs to be separated and categorized before being able to process it with various data science and machine learning algorithms and techniques. This is yet another Big Data challenge that we are going to continue facing for years to come. Remember that most algorithms need to work with a specific data format, therefore data exploration and analysis becomes a great deal more important and more time consuming.

Another issue is the fact that all the gathered information needs to be valued in some way. Just think of social media networks like Twitter. Imagine having to analyze all of the data ever recorded, all of the tweets that have been written since the conception of the platform. This would be extremely time consuming no matter the processing power of the computers managing it. All of the collected data would have to be pre-processed and analyzed in order to determine which data is valuable and in good condition. Furthermore, Big Data like what we just discussed raises the issue of security. Again, think about social media platforms, which are famous for data gathering. Some of the data includes personal information that belongs to the users and we all know what a disaster a Big Data breach can be. Just think of the recent Cambridge Analytica scandal. Another example is the European Union's reaction to the cybersecurity threats involving Big Data which led to the creation of the General Data Protection Regulation with the purpose of defining a set of rules regarding data gathering. Company data leaks can be damaging, but Big Data leaks lead to international disasters and changes in governments. But enough about the fear and paranoia that surrounds today's Big Data. Let's discuss the quality and accuracy of the information, which is what primarily concerns us.

A Big Data collection never contains a consistent level of quality and value when it comes to information. Massive datasets may contain accurate and valuable data that we can use, however, without a doubt it also involve a number of factors that lead to inaccuracy. One of the first questions you need to ask yourself is regarding those who have recorded the data and prepared the datasets. Have they made some assumptions in order to fill in the blanks? Have they always recorded nothing but facts and accurate data? Furthermore, you need to concern yourself with the type of storage system that was used to hold on to that data and who had access to it. Did someone do anything to change the data? Was the storage system damaged in some way that led to the corruption of a large number of files? In addition, you need to consider the way that data was measured. Imagine four devices being used in the same area to measure the temperature of the air. All of that data was recorded, but every device shows a different value. Which one of them holds the right values? Which one has inaccurate measurements? Did someone make any mistakes during the data gathering process?

Big Data poses many challenges and raises many questions. There are many variables that influence the quality and the value of data and we need to consider them before getting to the actual data. We have to deal with the limitations of technology, the possibility of human error, faulty equipment, badly written algorithms, and so on. This is why Big Data became a specialization of its own. It is highly complex and that is why we are taking this section of the book to discuss the fundamentals of Big Data and challenges you would face as a data scientist if you choose to specialize in it.

Applications in the Real World

Big Data is a component of Data Science, which means that as long as something generates and records data, this field will continue being developed. Therefore if you are still having doubts regarding your newly acquired skill, you should stop. Just think about a market crash. Imagine it as bad as any other financial disasters in the last century. This event would generate a great deal of data, personal, commercial, scientific and so on. Someone will have to process and analyze everything and that would take years no matter how many data scientists you would have available.

However, catastrophes aside, you will still have to rely on a number of pre-processing, processing and analysis in order to work with the data. The only difference is that datasets will continue to grow and in the foreseeable future we will no longer deal with small datasets like the ones we are using in this book for practice. Big Data is the future, and you will have to implement more powerful learning models and even combine them for maximum prediction accuracy. With that being said, let's explore the uses of Big Data in order to understand where you would apply your skills:

1. Maintenance: Sounds boring, but with the automation of everything, massive amounts of data are produced and with it we can determine when a certain hardware component or tool will reach its breaking point. Maintenance is part of every industry whether we're talking about manufacturing steel nails or airplanes. Big Data recorded in such industries will contain data on all the materials that were used and the various attributes that describe them. We can process this information and achieve a result that will immediately tell us when a component or tool should expire or need maintenance. This is a simple example of how Big Data analysis and data science in general can be useful to a business.
2. Sales: Think of all the online platforms and shops that offer products or services. More and more of the turn up

every single day. Large businesses are even warring with each other over the acquisition of data so that they can better predict the spending habits of their customers in order to learn how to attract those who aren't interested or how to convince the current ones to make more purchases. Market information is generated at a staggering pace and it allows us to predict certain human behavioral patterns that can generate more business and the improvement of various products and services.

3. Optimization: Big Data is difficult and costly to process and analyze, however, it is more than worth it since corporations are investing more and more into data scientists and machine learners. This kind of data analysis provides them with the much needed information to make their business more efficient and well-optimized. The larger a company is, the more cracks can be found through which they lose time and money. Predicting various costs and market trends can have a massive impact on a company's or even government's spending habits.

As you can see, analyzing Big Data is one of the next steps in data science as more and more data becomes available. Try to visualize the fact that the global society has generated nearly 90% of all information ever recorded in only two years. Every bit of data is valuable to some degree and you will have to learn all the algorithms and techniques that will help you process Big Data. With that being said, let's take a brief look at some of these techniques that are in use today to explore and process vast amounts of information.

Analyzing Big Data

In the previous chapters we have discussed a number of supervised and unsupervised learning algorithms that are used to explore, process, and analyze data. Additionally, we have also explored neural networks with the purpose of creating powerful models that can deliver accurate results. However, keep in mind that so far we talked about regular datasets. As previously mentioned, Big Data is different. Fortunately, some of the techniques and algorithms you learned can also be used to classify Big Data. In this section we will briefly summarize which methods can be used to process massive volumes of data.

When it comes to supervised learning algorithms you can use support vector machines, the Naïve Bayes algorithm, Boosting algorithm, the Maximum Entropy Method and more. Some of these are beyond a beginner's scope, therefore we did not focus on them in this book. However, you still have some options. Furthermore, you can also use a number of classification techniques such as the K-nearest neighbor algorithm or Decision Trees. Remember that classification simply involves discrete class attributes, while regression involves continuous class attributes. To solve Big Data regression problems you can use both the linear and logistic regression methods.

On the other side we have unsupervised learning algorithms which use unlabeled data in order to classify it by performing a comparison operation on all the data features. In this case you can rely on clustering algorithms, such as the K-means clustering algorithm we discussed earlier. Other techniques you can use are the Adaptive Resonance Theory and the Self Organizing Maps. Feel free to explore these last two on your own. Furthermore, you pursue clustering techniques at a deeper level because they are excellent when working with Big Data. For instance, you can use supervised clustering methods, which involve determining clusters that have a high probability density regarding the individual classes. Ideally, you want to use supervised clustering algorithms only when you have target variables and training sets which included them inside a cluster.

Another option is using unsupervised clustering techniques because they can lower the value of the intercluster similarity and therefore increase it. Keep in mind that these methods are best used on a specific function. K-means clustering is one of these techniques and it remains one of the most popular ones in use even when it comes to Big Data.

Finally, we have semi-supervised clustering algorithms. In this case we have more than just a similarity parameter. This category of clustering methods allows us to make modifications to the domain information. This way clustering can be significantly improved. This information can also act as limitations between the observations variables.

As a beginner, if you attempt working with Big Data, you should focus on using binary classifiers such as the support vector machines you learned about, or data classifiers such as decision trees. Stick to what is familiar for now, however, you should always feel encouraged to experiment on your own. As you can see there are many algorithms and techniques and we can fit only so much in one book. Feel free to use the knowledge you gained so far because it is all you need to get started on your own.

Summary

This chapter was a brief summary of what Big Data involves. You learned the history and fundamentals of this concept, as well as why you should consider focusing more on this topic. Data is generated all around us and soon, smaller datasets will be a thing of the past. Remember that information is generated at an exponential rate and currently our computers and the current number of data scientists and machine learners cannot catch up. There are even Big Data sets that have been gathered over a span of 20 years and they are still waiting to be processed and analyzed. This is a specialization of the future and you should seek to use what you have learned so far on such massive volumes of data. Some of the algorithms you studied can be applied as they are, while some need to be adapted. Use the toolkit you gathered so far and explore how it is to work with Big Data.

Conclusion

Congratulations are in order, as you successfully completed your fundamental training in data science. You came a long way from learning how to install your Python work environment. Sit back and allow time to solidify your knowledge because you have just processed a great deal of information. Data science isn't a light topic, and some of the techniques and algorithms you explored can be challenging.

Now that you have finished learning the basics, you should take the time to go over this book one more time to make sure you didn't overlook anything. It is vital that you understand the theory behind these learning algorithms and analysis techniques before you advance to the next stage. Take what you have learned so far, and make sure to practice every concept on your own with one of the widely available open source datasets. By learning everything in a structured manner and applying it into practice, you will become a data scientist in no time!

Bibliography

Albon, C. (2018). *Machine learning with Python cookbook: practical solutions from preprocessing to deep learning.* Sebastopol, CA: OReilly Media.

Garreta Raúl, & Moncecchi, G. (2013). *Learning scikit-learn: machine learning in Python: experience the benefits of machine learning techniques by applying them to real-world problems using Python and the open source scikit-learn library.* Birmingham, UK: Packt Publishing Ltd.

Géron Aurélien. (2019). *Hands-on machine learning with Scikit-Learn and TensorFlow: concepts, tools, and techniques to build intelligent systems.* Beijing ; Boston ; Farnham ; Sebastopol ; Tokyo: OReilly Media.

Zaccone, G., & Karim, M. R. (2018). *Deep Learning with TensorFlow: Explore neural networks and build intelligent systems with Python, 2nd Edition.* Birmingham: Packt Publishing.

Lightning Source UK Ltd.
Milton Keynes UK
UKHW021840011020
370869UK00013B/195